OIL ON THE SEA

OCEAN TECHNOLOGY

Series edited by John P. Craven

Department of Naval Architecture and Marine Engineering
Massachusetts Institute of Technology

OIL ON THE SEA

Proceedings of a symposium on the scientific and engineering aspects of oil pollution of the sea, sponsored
by Massachusetts Institute of Technology and Woods
Hole Oceanographic Institution and held at Cambridge,
Massachusetts, May 16, 1969

Edited by
DAVID P. HOULT

Department of Mechanical Engineering
Massachusetts Institute of Technology
Cambridge, Massachusetts

℗ SPRINGER SCIENCE+BUSINESS MEDIA, LLC 1969

First Printing – November 1969
Second Printing – December 1970
Third Printing – April 1972

Library of Congress Catalog Card Number 76-98411

ISBN 978-1-4684-9021-3   ISBN 978-1-4684-9019-0 (eBook)
DOI 10.1007/978-1-4684-9019-0

© 1969 Springer Science+Business Media New York
Originally published by Plenum Press, New York in 1969
Softcover reprint of the hardcover 1st edition 1969

FOREWORD

In the last decade, changes in the scale of operations required to find and transport oil have led to a pollution problem of major proportions: oil on the sea. These changes occurred slowly, and the change in magnitude of the possibilities for pollution went unrecognized until a series of dramatic accidents recently gave the problem wide-spread public notice. The Torrey Canyon and Santa Barbara episodes are discussed in this volume.

The changes in the scale of oil operations stem from an ever increasing demand for energy. In response to this demand, oil drilling from offshore rigs on the continental shelf has been rapidly developed. To inexpensively transport oil to the consumers of energy, huge supertankers, of ever increasing size, are being constructed. These ships effect economic savings at the expense of being relatively underpowered, and less maneuverable. Having very deep draft, they are constrained to operate on the high seas and the few deep harbors of the world. Every year there is more oil pumped from the sea floor. Every year more oil is transported over the sea. Approximately one tenth of one per cent of this oil each year is spilled on the sea.

The purpose of the present volume is to provide a summary of our current understanding of the problem of oil on the sea. Before describing in detail the topics presented, it seems well to point out that a major difficulty in fully understanding this problem is the diversity of disciplines required to describe the situation. An attempt is made here to cover the main features of oil on the sea, but unfortunately, the evolving technology of offshore drilling for oil is not covered.

The main feature of this diversity can be seen by surveying the table of contents. The first three articles deal with the biological effects of oil spills. The next four articles are discussions of various engineering problems which arise in attempting to deal with spilled oil. Then there is an article which describes the development of supertankers for transporting oil, followed by the last article, on the legal and legislative government attempts to cope with this problem.

Clearly, these topics are not unrelated to one another. For example, the government can only enforce control procedures if these procedures are technologically feasible. The type of biological damage caused by an oil slick should be considered in choosing between various enforcement procedures. Again, the detergent used in cleaning up the Torrey Canyon spill was biologically more destructive than the oil spilled. Similar errors should be prevented in the future by attempting to gain a better understanding of all the facets of oil pollution.

Until very recently, these problems attracted very little attention. However, due to the Torrey Canyon and the Santa Barbara episodes, federal funds for research and development of various methods dealing with oil spills are becoming available. Thus, the technological state of affairs described in this book may be expected to change in the future. For this reason, the volume is not complete in itself. It is rather a status report of current efforts. Perhaps it will also serve as a guide for those who are working to provide for a cleaner ocean.

ACKNOWLEDGEMENT

I would like to acknowledge the work of James A. Fay, Professor of Mechanical Engineering at the Massachusetts Institute of Technology, who organized the symposium out of which this volume grew. Without his efforts, and the support of M.I.T. and the Woods Hole Oceanographic Institution, this volume would not have been possible.

David P. Hoult
Cambridge, Massachusetts
September 7, 1969

CONTENTS

EFFECTS OF 'TORREY CANYON' POLLUTION ON MARINE LIFE

Norman A. Holme

Marine Biological Association of the United
Kingdom
The Laboratory, Citadel Hill
Plymouth, Devon., England

The 'Torrey Canyon' stranded on the Seven Stones Reef off Lands End on March 18th, 1967. Within ten days her entire cargoe of 118,000 long tons of Kuwait crude oil had been released on to the sea, or burned following the bombing of the wreck. The released oil drifted off in three different masses; initially about 30,000 tons escaped and this drifted up the English Channel and polluted the N. coast of France, also Guernsey. During the following week a further 20,000 tons escaped, to pollute the west Cornish coast. On March 26th, the ship broke her back and released 50,000 tons which drifted south into Biscay. The wreck was bombed on March 28-30, the remaining 20,000 tons being either burned or released on to the sea.

The direction and speed of drift of oil on the sea surface can be computed from meteorological data--drift being in the same direction and at 3.4% of the wind speed. Provided winds are fairly strong, tidal currents can be ignored in the calculations. With this knowledge reconnoitering ships and planes can be sent to calculated positions rather than having to search at random.

Oil released on the sea surface changes in properties: water-soluble toxic substances are released so that the oil becomes virtually non-toxic, the lighter fractions evaporate so that the residues become progressively more difficult to burn, and a water-in-oil emulsion, containing up to 80% water, called 'chocolate mousse' may be formed.

1

Off the Cornish coast the oil at sea was sprayed both at sea and on the shores by detergents, the most commonly used being BP1002. Emulsification was often incomplete, and oil seems to have often reseparated from the unstable emulsions formed. These detergents were highly toxic, in concentrations of 10 p.p.m. or less, to many marine animals and plants. The most noticeable effect was the killing of limpets which normally browse over the intertidal rocks, with a resultant prodigious growth of green weed. Offshore, toxic effects were noticed off some heavily sprayed beaches, but in general mortalities were confined to the shore and there were no effects on plankton and fisheries. The toxic aromatic solvent fraction of the detergent was found to be volatile so that there was no progressive buildup of toxicity.

Most of the polluted Cornish coast was detergent-sprayed, but there was little damage to marine life on the few polluted stretches which had not been sprayed. Although lethal to sea birds, the oil only killed shore animals and plants if it was so thick as to smother them. There is evidence that limpets browse over the rocks, helping to re-move the oil.

The residue of 30,000 tons of oil released after the initial stranding arrived on the N. coast of Brittany some three weeks later, where it caused very heavy pollution along about 40 miles of coast between Trébeurden and Ile de Brehat. Because of the important inshore shellfisheries not much detergent was employed and manual methods of clear-ance using straw or gorse to soak up the oil, pumping and bailing of the oil, and bulldozing of sandy beaches were employed. Because of the dissected nature of the coastline it was not possible to clean up the whole coast, and it is now possible to see in many places what happens to oil if it is left untreated. On certain sandy beaches there was still a thick coating of oil on the surface 15 months after the original pollution. On heavily polluted rocks between tide marks the oil was beginning to crack and flake off the rocks, but at high tide mark it seemed to be firmly baked on the rocks. It will be several years before these rocks are clean.

The residue of 50,000 tons released when 'Torrey Canyon' broke her back drifted south into Biscay, where it remained at sea for two months, finally causing only light pollution on the west coast of Brittany. During this time about 50%

of the lighter fractions would have evaporated. Meanwhile the French had been treating the oil with natural chalk ($CaCO_3$) with an additive of stearic acid which made the chalk oleophilic and hydrophobic. About 3,000 tons of chalk were sprinkled on the oil, and it is believed that this caused the greater part of it to sink or disperse. There have been no subsequent reports that the sunken oil fouled fishing gear on the sea bed or was subsequently washed up on the beach. The success of this operation was due to the higher density of the floating oil after lighter components had evaporated, the length of time the oil stayed at sea where it could be treated, and the comporatively calm weather during this period.

REFERENCES

1. Cabinet Office (1967): The Torrey Canyon. Her Majesty's Stationery Office, London. 48 pp. (S.O. Code No. 63-190. Price 5s.9d).

2. Smith, J.E., ed. (1968). 'Torrey Canyon' Pollution and Marine Life. Cambridge University Press. 196 pp. Library of Congress Catalogue No. 68-21400. $9.50 in U.S.A.

OIL POLLUTION OF THE OCEAN[*]

Max Blumer

Woods Hole Oceanographic Institution

Woods Hole, Massachusetts   02543

[*]Contribution Number 2336 of the Woods Hole
Oceanographic Institution

THE EXTENT OF MARINE OIL POLLUTION

Oil pollution is the almost inevitable consequence of
the dependence of a rapidly growing population on a largely
oil-based technology. The oil reserves which have accumu-
lated in the earth during the last 500 million years are
being depleted rapidly and will be exhausted within a few
hundred years. The use of oil or of other natural resources
without losses is impossible; losses occur in production,
transportation, refining and use. The immediate effects of
large scale spills in coastal areas are well known but only
through the recent introduction of marine surface sampling
tools have we become aware of the degree of oil pollution
of the open ocean. Thus, during a recent cruise of R/V
CHAIN of the Woods Hole Oceanographic Institution to the
southern Sargasso Sea, many surface "Neuston" net hauls were
made to collect surface marine organisms. These tows were
made between 32°N - 23°N latitude (corresponding to a distance of 630 miles) at a longitude of 67°W. Inevitably,
during each tow, quantities of oil-tar lumps, up to 3 inches
in diameter were caught in the nets. After 2 - 4 hours of

Our work on the fate of hydrocarbons in the sea has long been
supported by ONR (present grant: NOO 14-66-contract CO-241)
and by NSF (present grant: GA-1625). Present work is also
supported by a grant from FWPCA (18050 EBN).

towing, the mesh became so encrusted with oil that it was
necessary to clean the nets with a strong solvent. On the
evening of 5 December 1968, between 1835 - 2240 R hours at
25°40'N, 67°30'W, the nets were so fouled with oil and tar
material that towing had to be discontinued. It was estimated that there was 3 times as much tar-like material as
Sargasso Weed (on a volume basis) in the nets[1]. Similar
occurrences have been reported worldwide by observers from
this as well as from other Institutions.

In order to arrive at a figure for the total oil influx
into the ocean from various sources, we need figures for the
total amount of oil produced, shipped and for the fraction
lost in shipping and handling. The world oil production
stands near 1.8×10^{13} g/year. Of this amount at least 60%
or 10^{15} g/year is transported across the ocean. Much of the
transport is concentrated in restricted shipping lanes;
thus, 25% of the world production passes through the
English Channel!

A minimum estimate of the fraction of oil lost can be
calculated from the extent of single large accidents and
from operating records of oil ports. Thus, the tanker,
Torrey Canyon, alone carried and lost 10^{11} g or 0.01% of the
annual oil transport across the sea. The recent accident
at Santa Barbara has introduced into the ocean 10^{10} g of
crude oil. Reliable figures about oil losses in port are
available from Milford Haven, a relatively new oil English
port, adjacent to a national park. There, great efforts have
been made to control and prevent oil pollution and to keep
a record of the size of any spills. In 1966 the annual
turnover at Milford Haven was 3×10^{13} g. The losses in the
same time period amounted to 2.9×10^{9} g or 0.01% of the total
amount handled. A single accident (the tanker, Chrissi P.
Goulandris) contributed between 10 and 20% of this total[2];
the other losses are attributed to design faults, breakages,
and mechanical failures, losses in transfer and human error[3].
This figure does not include losses outside the port due to
accidents in shipping (e.g. the Torrey Canyon) and from
numerous other sources such as ballasting and flushing of
the bilges, etc. With the less stringent operation of many
other ports and the additional losses on the high sea, the
loss in transport alone may amount 0.1% of the total oil
shipped. The actual oil influx to the ocean is higher, since
the figures above do not include accidents in production
(e.g. Santa Barbara) return to the ocean of petroleum

products (fuels and spent lubricants) in untreated municipal wastes and incomplete combustion of marine fuels.

Therefore, the oil influx to the ocean from shipping losses only is about 10^{12} g/year; other causes like influx from sewage and incomplete combustion may add substantially higher amounts.

OIL COMPOSITION AND BIOLOGICAL EFFECTS

In order to assess the biological effects of the oil pollution we should discuss the composition of crude oil and the relative toxicity of its fractions. Crude oil is one of the most complex mixtures of natural products, extending over a very wide range of molecular weights and structures (Figure 1). The low boiling saturated hydrocarbons have, until quite recently, been considered harmless to the marine environment. However, it has now been demonstrated that these hydrocarbons produce at low concentrations anaesthesia and narcosis and at greater concentration cell damage and death in a wide variety of lower animals and that they may be especially damaging to the larval and other young forms of marine life[4]. Higher boiling saturated hydrocarbons naturally occur in many marine organisms and are, probably, not directly toxic though they may interfere with nutrition and possibly with the reception of the chemical clues which are necessary for communication between many marine animals. Olefinic hydrocarbons probably are absent from crude oil, but they are abundant in oil products, e.g. in gasoline and in cracking products. They are also produced by many marine organisms, and may serve biological functions, e.g. in communication. However, their biological role is poorly understood. Aromatic hydrocarbons are abundant in petroleum; they represent its most dangerous fraction. Low boiling aromatics (benzene, toluene, xylenes, etc.) are acute poisons for man as well as for all other organisms. It was the great tragedy of the Torrey Canyon accident, that the detergents which were then used to disperse the oil spill had been dissolved in low boiling aromatics. Their application multiplied the damage to coastal organisms. It should be pointed out, however, that poisoning of marine life will occur even with non-toxic detergents or dispersants which are applied in non-toxic solvents, because they disperse the toxic materials of crude oil. This exposes organisms to these poisons through contact and ingestion. They high boiling aromatic hydrocarbons

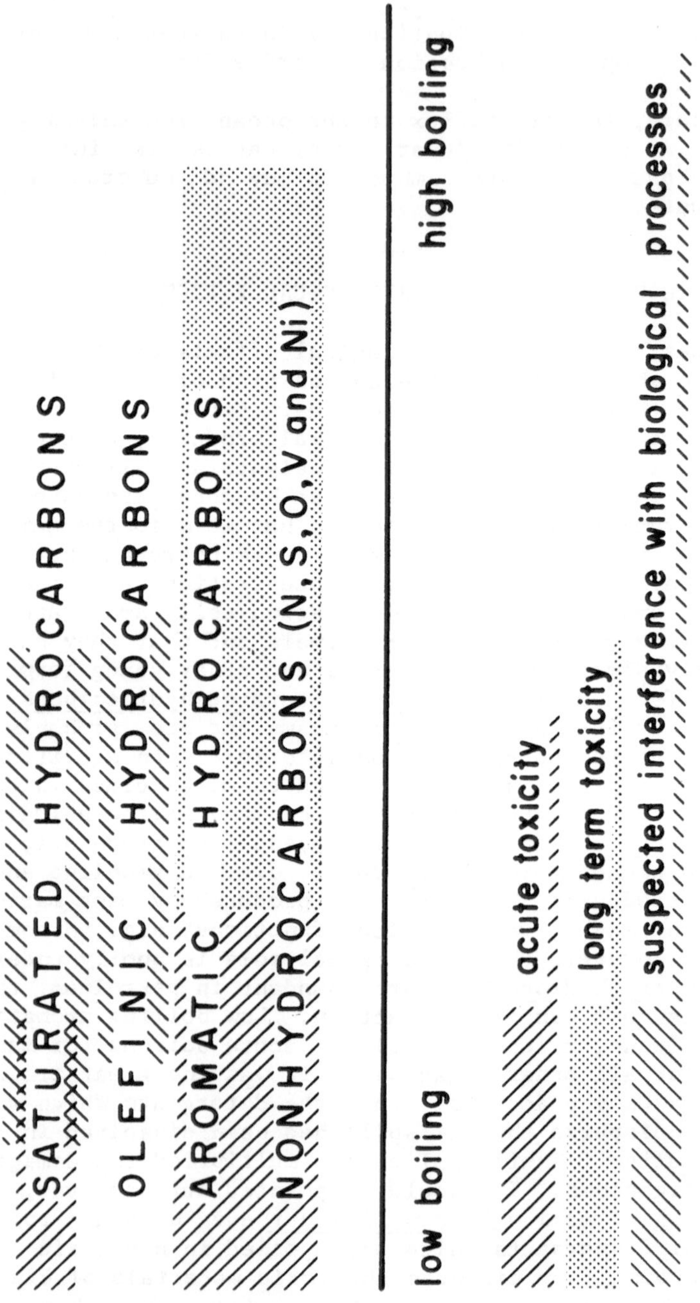

Figure 1. Composition of Crude Oil and Toxicity of its Fractions

are suspected as long term poisons. Current research on the carcinogenic hydrocarbons in tobacco smoke has demonstrated, that the carcinogenic activity is not--as was previously thought--limited to the well known 3.4-benzopyrene. A wider range of alkylated 4- and 5-ring aromatic hydrocarbons can act as potent tumor initiators[5]. While the direct carcinogeneity of crude oil and crude oil residues has not yet been conclusively demonstrated, it should be pointed out that oil and residues contain alkylated 4- and 5-ring aromatic hydrocarbons similar to those in tobacco tar. In their behavior and toxicity the nonhydrocarbons of crude oil (nitrogen, oxygen, sulfur and metal compounds) closely resemble the corresponding aromatic compounds.

OIL ANALYSIS AND LAW ENFORCEMENT

The great complexity of crude oil has an interesting consequence: The variety in the composition of different crude oils and oil products is so great that every oil has its own compositional features which are typical and persistent like a fingerprint. Great efforts have been expanded by many oil companies in utilizing this characteristic for correlating or distinguishing oils produced from different oil bearing horizons or for correlating oils with their source sediments. This fingerprinting technique is now becoming available to the public and will lead to improved and often conclusive correlation of an oil spill with oil from a particular oil field or from a particular vessel[6],[7] The analytical techniques are simple and should be a great aid to law enforcement.

LONG TERM EFFECTS OF OIL POLLUTION

The immediate, short term effects of oil pollution are obvious and well understood in kind if not in extent. The coastal fouling and damage to bird populations has been documented abundantly. As mentioned above, fouling on the high seas is just now being recognized, even though the amount of tar at the sea surface already exceeds the amount of surface plant life. The short term toxicity has been discussed above for individual petroleum fractions. In contrast to this, we are rather ignorant about long term and low level effects of crude oil pollution. I fear that these may well be far more serious and long lasting than the more

obvious short term effects. I wish to discuss long term
toxicity and low level interference of oil pollution with
the marine ecology.

    In combination, the great complexity of the marine food
chain and the stability of the hydrocarbons in marine
organisms, lead to a potentially dangerous situation. The
food chain of those <u>terrestrial</u> organisms, which are important for human nutrition, is simple. Man either eats plant
material or meat products from animals that have been raised
on plant food. <u>Human food</u> derived <u>from the sea</u> is much more
remote from its origin in plants. Few marine plants are
directly used for human nutrition and, except for shellfish,
we consume few marine animals that have fed directly on
marine plants. Most larger marine animals derive their food
from other marine animals that are already remote from the
original plant source. We have studied the fate of organic
compounds in the marine food chain and have found that hydrocarbons, once they are incorporated into a particular marine
organism, are stable, regardless of their structure, and
that they may pass through many members of the marine food
chain without alteration[8],[9]. In fact, the stability of
the hydrocarbons in marine organisms is so great that hydrocarbon analysis serves as a tool for the study of the food
sources of marine organisms. In the marine food chain hydrocarbons may not only be retained but they can actually be
concentrated. This is a situation akin to that of the
chlorinated pesticides which are as refractory as the hydrocarbons. These pesticides are concentrated in the marine
food chain to the point where toxic levels may be reached.
It is likely that the treatment of oil spills with detergents or dispersants, or the natural dispersion of oil in
storms produces oil droplets of a particle size range that
is ingested and assimilated by many marine organisms. Once
assimilated, this oil passes through the marine food chain,
and eventually reaches organisms that are harvested for
human consumption. One consequence will be the incoporation
into food of materials which produce an undesirable flavor.
A far more serious effect is the potential accumulation in
human food of long term poisons derived from crude oil, for
instance of carcinogenic compounds.

    Another concern is the possible long term damage by
pollution to the marine ecology. Many biological processes
which are important for the survival of marine organisms
and which occupy key positions in their life processes are

mediated by extremely low concentration of chemical messengers
in the sea water. We have demonstrated that marine predators
are attracted to their prey by organic compounds at concentrations below the part per billion level[10]. Such chemical attraction--and in a similar way repulsion--plays a role
in the finding of food, the escape from predators, in homing
of many commercially important species of fishes, in the
selection of habitats and in sex attraction. There is good
reason to believe that pollution interferes with these processes in two ways: by blocking the taste receptors and by
mimicking for natural stimuli; the latter leads to false
responses. Those crude oil fractions likely to interfere
with such processes are the high boiling saturated and
aromatic hydrocarbons and the full range of the olefinic
hydrocarbons. It is obvious that a very simple--and seem-
ingly innocuous--interference at extremely low concentration
level may have a disastrous effect on the survival of any
marine species and on many other species to which it is tied
by the marine food chain.

COUNTERMEASURES AGAINST LARGE OIL SPILLS

It must be obvious from this discussion that I do not
consider the use of detergents or dispersants, toxic or non-
toxic, as a solution for pollution problems. The introduction by dispersants of the toxic components of crude oil into
the sea and the marine food chain constitutes a risk that
should not be taken lightly.

Sinking of an oil spill by treatment with hydrophobic
minerals (e.g. chalk treated with stearic acid or refrac-
tories treated with silicones) may be preferred; however,
we do not know whether the oil remains on the sea floor or
whether it will return to intermediate or shallow waters
where it can enter the food chain. Also, we do not know
enough about the effect of oil on bottom communities. Sedi-
mentation rates in the open ocean are quite low, and oil
that has been sunk will remain exposed for very long periods
of time. In my opinion, burning of the oil where possible
or containment and rapid recovery are the only acceptable
solutions for managing large spills.

THE LONG-TERM OUTLOOK

Mankind is depleting the natural oil reserves rapidly. Therefore, it is unlikely that oceanic oil transport will increase by several orders of magnitude. In spite of this there are several good reasons to anticipate an increase in the seriousness of the marine oil pollution. Marine oil transport through more hazardous waters will increase (e.g. transport of the Alaskan oil through the Bering Straits). Oil production will shift increasingly to the continental shelves and oil reserves in very deep water (e.g. Sigsbee Deep, Gulf of Mexico) may be tapped. Both will lead to an increasing risk of accidents. Oil products and synthetic oil (coal hydrogenation products, shale oil), which are more toxic than crude oil, will make up a larger fraction of the oil transported, used and spilled.

We are convinced of the great value of oceanic food production for mankind. In the future, a larger fraction of human nutrition must be derived from the sea. Farming of the sea (aquaculture) will become an important pursuit for man. However, if we do not take care of the present biological resources in the sea, we may do irreversible damage to many organisms, to the marine food chain and we may eventually destroy the yield and the value of the food which we hope to recover from the sea.

REFERENCES

1. V. E. Noshkin and J. E. Craddock, Event Information Report, Smithsonian Institution. 17 December 1968, Event #66-68.

2. D. R. Arthur, The Biological Problems of Littoral Pollution by Oil and Emulsifiers - A Summing Up, Field Study Council, Suppl. Vol. 2, (1968), 159.

3. G. Dudley, The Problem of Oil Pollution in a Major Oil Port, Field Study Council, Suppl. Vol. 2, (1968), 21.

4. R. J. Goldacre, Effects of Detergents and Oils on the Cell Membrane, Field Study Council, Suppl. Vol. 2, (1968), 131.

5. E. L. Wynder and D. Hoffman, Experimental Tobacco Carcinogenesis, Science, 162, (1968), 862.

6. J. V. Brunnock, D. F. Duckworth and G. G. Stephens, Analysis of Beach Pollutants, J. Inst. Petroleum, 54, (1968), 310.

7. S. J. Ramsdale and R. E. Wilkinson, Identification of Petroleum Sources of Beach Pollution by Gas Liquid Chromatography, J. Inst. Petroleum, 54, (1968), 327.

8. M. Blumer, Hydrocarbons in Digestive Tract and Liver of a Basking Shark, Science, 156, (1967), 390.

9. M. Blumer, "Zamene", Isomeric C_{19} Monoolefins from Marine Zooplankton, Fishes and Mammals, Science, 148, (1965), 370.

10. K. J. Whittle and M. Blumer, Chemotaxis in Starfish, Symposium on Organic Chemistry of Natural Waters, University of Alaska, Fairbanks, Alaska, 1968 (in press).

THE SANTA BARBARA OIL SPILL

Robert W. Holmes

University of California

Santa Barbara, California

My task this morning is to discuss with you the recent Santa Barbara oil spill. To present an unbiased and complete picture of what has been and is happening in Santa Barbara is virtually impossible due to the complexity of the event and the large number of organizations and individuals involved in responding to the disaster. In recent weeks information sources have been drying up, presumably because of the pending litigation. Nevertheless, I will present the facts and opinions as I have been able to learn, admittedly incomplete in certain areas.

My sources of information have been varied and I would at this time like to acknowledge these. Drs. Joseph Connell, Alfred Ebeling, and Michael Neushul of the Biological Sciences Department of UCSB and Dr. Paul Smith of the Bureau of Commercial Fisheries (La Jolla) have provided useful information on the biological effects of the oil contamination. Together with Mr. John Cubit, Floyd DeWitt, Jr., Michael Foster, and Don Potts, all graduate students at UCSB, these individuals have spent a great deal of time and effort in monitoring the biological effects of the spill. In this, a number of these individuals have been aided by financial support from the F.W.P.C.A. (Federal Water Pollution Control Board). Mr. Al Allen of the General Research Corporation of Santa Barbara has provided his estimates

of the amount of oil seeping from the vicinity of Platform
A. The Santa Barbara News Press has provided intensive
coverage of the oil disaster and has been used as an
occasional source of information. The News Press has not
been completely objective in all of its reporting and is
therefore not always a reliable source of information. The
information presented on the response of the U. S. Coast
Guard to the disaster was taken from a tape recording of
remarks by Lieutenant George Brown, on the scene commander,
at a symposium held at UCSB and sponsored jointly by UCSB
and the Science and Technology Society of Santa Barbara.
The First and Second Field reports issued by the U. S.
Geological Survey have been quoted extensively to provide
you with a detailed account of what actually happened on
Platform A. Lastly, my own observations taken during
flights in the area have been used on occasion.

The well which was responsible for the oil disaster
is located on Union Oil's Platform A located approximately
$5\frac{1}{2}$ miles south of Santa Barbara on Federal lease OSC-P-0241.
Four companies share this lease: Union, Texaco, Gulf, and
Mobile. Well 21, which was responsible for the oil spillage
was the fifth well drilled on Platform A. Three of four
wells had been completed but were not in production on
January 28, 1969. The fourth well reached a depth of 3030
feet on January 29, 1969, but had not been completed for
production at the time the difficulties with Well 21 began.
No difficulties were encountered with these four wells.

"Drilling of the well ((No. 21)) was commenced
on January 14, 1969. ((The)) well ((had been)) drilled
to total depth of 3479 feet on 28 January. The 13-3/8"
casing was set at 514 feet and cemented up to the
surface of the ocean floor. The bottom of the
casing is 238 feet in rock below the ocean floor.
No additional casing was set below this depth.
If the well had been completed for production, a
production string of casing was proposed to be set
at total depth and cemented in place.

"At 10:45 a.m., on Tuesday, January 28, after

Figure 1. Platform A, showing the oil boiling up from well 21. A dispersant is being sprayed on the oil by the boat in the lower right-hand corner.

Figure 2. A beach near the oil spill, showing a heavy coating of oil.

drilling to total depth, the drill pipe was being pulled from the hole to run electric logs when mud began to flow from the drill pipe. Eight stands of pipe, approximately 720 feet, had been removed when the mud began to flow. The rig crew made unsuccessful attempts to control the flow by installing a threaded valve, and then the Kelly on top of the drill pipe. Within a few minutes it was decided to drop the drill pipe in the hole and close the blind rams on the blowout preventer. This action effectively controlled the flow of mud and gas from the casing of the well.

"Soon after the flow from the wellhead was contained, gas and some oil began to boil up through the water near the platform. Initially the largest boil was about 800 feet east of the platform. On the next day all of the boils had completely subsided except a single boil of gas and oil about 30 feet in diameter, the center located about 20 feet from the northeast leg of the platform. It appears that a small fault outcrops on the ocean floor near the platform. The location of the outcrop and the strike of the fault appears to coincide with the location of all of the boils which had been observed. Union Oil Company officials expressed the probability that the gas and oil escaping from the ocean floor was flowing up the hole, laterally through porous strata, and then to the ocean floor through the small shallow fault. The Union Oil Company first reported the trouble on Well A-21 to the Los Angeles office of the Geological Survey at 1:00 p.m. Tuesday, January 28 . . .

"The initial attempt to control the flow from the ocean floor was made by running drill pipe into the hole through special wellhead equipment and stabbing and reconnecting with the top of the fish (string of drill pipe which had been dropped into the hold Tuesday). Reconnection was successful but it was found upon attempting to circulate mud that the bottom of the drill pipe was plugged and stuck. This

condition prevented the circulation of heavy mud
which was needed to stop the flow from the drill
hole. Several attempts were then made to back off
the drill pipe at a point below a back pressure valve
which had been installed at the bottom of the drill
pipe which had been run back into the hole. This
valve permits the flow of fluids down the drill pipe
but not back up the drill pipe. . .

"Being unable to remove the back pressure valve
from the drill pipe, it was decided to attempt to
mill out the moveable parts of the valve so that a
gun perforator could be lowered through the valve
to the bottom of the drill pipe. . . Actual milling
operations began early Monday morning and were completed in approximately $2\frac{1}{2}$ hours. The perforating
gun was run through the valve without difficulty to
2942'. The drill pipe was perforated from 2860
feet to 2883 feet with 97 perforating bullets (.34
inch diameter per hole). The circulation of sea
water was commenced but did not stop the flow of
oil and gas from the sea floor. Sea water was initially
used because it is not as easily gas cut as would be
an initial injection of drilling mud. Early on
Tuesday the circulation of drilling mud was commenced.
Approximately 3000 barrels of heavy weighted drilling
mud was pumped into the well bore through the drill
pipe and perforations over a period of 6 hours on
Tuesday. There was a definite reduction in the
amount of gas and oil blowing out and during short
time intervals the boil at the northeast corner of
the platform subsided. . . After 6 hours of pumping
and with diminishing supplies of the heavy specially
mixed liquid mud, it was determined that increased
pumping rates and increased volumes of supplies would
be necessary. It was decided to cease pumping mud
until additional mud pumping capacity could be lifted
on the platform deck and huge supplies of liquid mud
could be assembled for the massive injection dose
into the well bore.

"On Wednesday, January 29, the decision was made
to propose to drill a relief well as a backup measure
in the event operations on Well A-21 would be unsuccessful. ((Permission granted by USGS on 30 January)).
Drilling of the relief well ((1000 feet south of Well
21 to be drilled by a floating drilling rig)) was
commenced at 7:00 a.m. Sunday, February 2. The
relief well had been drilled to a depth of about 600
feet on Tuesday, February 4.

"By early Friday, February 7, the storm front
had moved through the area and at daylight Union
began moving the mud barges and other equipment out
of Santa Barbara harbor to the platform area. At
approximately 1:00 p.m. . . the pumping of sea water
weighted with calcium chloride, into the well was
commenced. This was followed by heavy mud.

". . . At 4:30 p.m., slight mud returns to the
ocean surface were observed from the platform . . .
At 8:00 p.m., on Friday there was no visible ((evidence))
of bubbles. At that time it was decided to plug the
entire well bore with cement. The pumping of mud at
slow rates continued through both the drill pipe and
casing annulus until the pumping of cement was commenced. . .

"By 12:00 midnight on Friday, 1150 sacks of
cement slurry had been pumped. . .

"On Saturday, February 8, Fred L. Hartley,
President of Union Oil Company, released the following
to the press. "In our opinion the well has been
plugged. As anticipated, minor residual gas is
continuing to work its way to the surface. For
added insurance the hole will be filled with cement
to the surface."

The above quotations were taken verbatim from the First
and Second Field Report of Union Oil Company Leases OCS-P-
0241, Platform A, Well No. 21, issued by the U. S. Department of the Interior, Geological Survey on February 4

and 9, 1969, respectively. The remarks contained within
double brackets were inserted by the present author for
sake of clarity. Portions of the report have not been
included--breaks in the sequence are indicated by three
periods.

The amount of casing required by the State of California on state leases differs from the amount required on
Federal leases. For wells less than 5000 feet in depth
the State of California requires 300 feet of conductor
casing and 1000 feet of surface casing below the ocean
floor to be cemented solid. Federal regulations require
300-500 feet of casing for wells which do not exceed 7000
feet in depth. Well 21 had only approximately 239 feet
of casing in the bottom and as nearly as I can determine,
Union Oil Company was granted a variance in drilling Well
21 since the casing was less than 300 feet. Naturally,
the question arises as to whether or not the blowout on
Well 21 could have been prevented by employing more
stringent requirements like those of the State of California. Opinions vary judging from the reports in the
press. The reaction of the Geological Survey to such
comments were as follows:

"Items in the press, attributed to officials of
the State of California, indicating that the State
well casing requirements for off-shore wells are
more strict than Federal requirements. We do not
agree with these statements, if made. It should be
clear to all concerned that both State and Federal
requirements which currently exist are merely guidelines which can, have been, and should be varied for
different areas where gas and oil exploration and
development is conducted. Where geologic and reservoir
engineering data are known, it would not in many cases,
be good engineering practice to adhere to the printed
guidelines." (From the Second Field Report of the U.S.
Geological Survey, February 9, 1969).

Judging from the recently revised drilling regulations
issued by the Interior Department and additional regulations

being considered, it would appear that stricter requirements
or guidelines could have prevented the Santa Barbara blowout.
Whether or not strict adherence to the state of California
casing requirements would have prevented the blowout is not
known to the public at this time.

During the period January 28 to midnight on February 7,
when oil leakage from Well 21 was uncontrolled, Union Oil
estimated the flow of oil into the Santa Barbara Channel
had occurred at a rate of 500 barrels per day or approximately 21,000 gallons per day--or a total of roughly
230,000 gallons. This is an order of magnitude less than
the minimum estimates made by Mr. A. Allen of the General
Research Corporation who suggests that the rate must have
been at least 5,000 barrels per day during this initial
period.

Mr. Hartley's announcement, quoted earlier, implying
that the leakage had been effectively stopped at midnight,
February 7, 1969, was premature for leakage of oil and gas
to the sea surface adjacent to Platform A still continues.
The rates of flow fluctuate but remain considerably less
than those first reported. Union Oil estimates the present
leakage to be about 10 barrels per day while Mr. Allen
estimates it to be roughly 200 barrels per day. Mr. Allen
estimated that by the 100th day of the spill, 3,250,000
gallons of oil had spilled into Channel waters. This is
roughly the amount of oil that reached the shores of Cornwall
after the Torrey Canyon disaster.

The causes of disparity in these two estimates remains
obscure since neither the Geological Survey nor Union Oil
have explained publicly the manner in which their estimates
of volume of flow have been obtained. Mr. Allen, however,
has discussed his methods and calculations openly and they
appear to be based on sound principles and considerations.

To date attempts to stop the spillage of oil into
the Santa Barbara Channel at Platform A have failed and
judging from press reports attempts to disperse the oil
by chemical means and to collect it by mechanical means

have not been particularly successful.

According to Federal regulations the Interior Department under the Oil Pollution Act of 1924 is responsible for oil spills. After the Torrey Canyon disaster the Departments of Interior and Transportation prepared a National Contingency Plan to deal with pollution of water, including oil pollution. This plan, approved by the President of the United States in November, 1968, involves the Departments of Interior, Transportation, Defense, and Health, Education, and Welfare. According to this plan the Interior Department has responsibility for dealing with pollution in fresh water and the Coast Guard has the responsibility in marine and brackish waters. In April and May, 1968, meetings were held in Santa Barbara to make plans for the initiation of the National Contingency Plan in case of an emergency. As mentioned above, the Coast Guard has the responsibility in cases of marine pollution.

Accordingly, at 12:50 p.m. on January 28, 1969, the local office of the Coast Guard was notified by Union Oil that they had experienced a blowout on Platform A at 10:45 a.m. Lt. George Brown, officer in charge in Santa Barbara Coast Guard detachment notified his superiors in Los Angeles of the blowout. At 6:10 p.m., Lt. Brown was notified by Union Oil that the leak was out of control. Lt. Brown requested an overflight of the area early on January 29. By 1:45 p.m. on the 29th he had notified all state, county, and city officials of the leak and after another overflight at 5:00 p.m. recommended that the disaster plan be implemented. At 9:00 p.m. on January 29, 1969, Lt. Brown held the first meeting with all concerned officials. Lt. Brown became "on the scene commander" charged with efforts to deal with the spill. Mr. Paul De Falco, from the San Francisco office of the F.W.P.C.A. was placed in charge of the clean-up operations.

It is my understanding that Union Oil was not required by law to follow any of the F.W.P.C.A.'s recommendations regarding clean-up procedures because the Federal government has no authority to deal with oil pollution caused by

offshore platforms on federal leases. The State of California has certain regulations dealing with oil pollution
caused by operations within the jurisdiction of the State.
In spite of the lack of Federal authority, Mr. De Falco's
recommendations were apparently followed by Union Oil.

Mr. DeFalco and the F.W.P.C.A. were painfully aware
of the toxic effects of detergents used in Great Britain
after the Torrey Canyon disaster and permitted the use of
chemicals only in certain areas and then only sparingly.
Corexit, a chemical dispersant, was used in the vicinity
of Platform A to help reduce the extreme fire hazard there.
Corexit was also employed in Santa Barbara Harbor which
received massive injections of oil and in the vicinity of
Anacapa Island. I have not been able to find any statements
concerning the effectiveness of Corexit in the Santa Barbara
spill. My own observations made from the air over Platform
A on two occasions suggest that the treatment was ineffective,
although it is possible that some reduction of fire hazard
in the vicinity of the platform was achieved. Other methods
to assist in the cleaning up of the sea surface were also
used. Talc and perhaps chalk, cement dust, and Corbalon
were tried and proved ineffective. Straw was used more
extensively than any other material. The straw was chopped
up and blown onto the sea surface from boats or barges by
threshing-machine-like devices. Straw absorbs oil readily
and proved relatively easy to pick up on the beaches once
it drifted ashore. Such methods relocate the oil rather
than remove it from the environment directly.

A number of devices have been employed by Union in
efforts to contain the oil flowing from the ocean floor
and/or to make removal of oil from the sea surface more
effective. The booms, skimmers, blankets, and funnels tried
in the vicinity of Platform A have not been notably successful and the need for more effective devices which will with-
stand the vicissitudes of the marine environment and
operate efficiently under a wide range of sea states is
recognized by everyone.

Intensive and extensive efforts to clean up the oil

which reaches the shore has been and is being made by Union's
clean-up crews. These procedures are mechanical and involve
removing oil-soaked sand and straw for disposal inland.
Because these crews only clean up the beaches after oil has
reached them and then often not promptly, layers of oil-
soaked sand may be found below the present sand surface.
These oil-contaminated sands will become exposed again next
year after winter storms. Oil-covered rocks above high tide
level exist between Pt. Conception and Ventura and on the
Channel Islands, some 70-110 miles of coast. Steam cleaning
of these has proved successful but the areas so treated are
quite limited in extent. This leaves miles of oil-covered
rocks adjacent to the beach visible to the passing motorist
and beach users. Such a vista destroys or impairs the
esthetic experience usually obtained in an unpolluted environment.

The effects of the oil disaster are difficult to
evaluate at the present time. The beaches were severely
contaminated with heavy crude and although the cleaning-
up procedures have been effective, oil still continues to
drift up on the beaches. Since cleaning operations take
place only after a beach has become oily, we are still faced
with the unpleasant sight of recurring contamination. The
economic impact of this upon the resort and tourist trade
is still unknown.

An exact assessment of the overall biological effect
of the oil spill is impossible at this time. A series of
unusual natural events occurred roughly simultaneously with
the Union spill which makes it difficult to relate present
or future biological change specifically to oil pollution.
For reasons unknown, the ocean surface temperature in the
Santa Barbara region is $2^{o}F$ colder than the ten year average.
According to Dr. A. Ebeling, UCSB ichthyologist, such
changes can have a pronounced effect upon the local commercial
and sport fishery. In addition excessive rains in January
and February diluted the tide pools and inshore surface
water appreciably for relatively long periods of time. Such
prolonged exposure to low salinity conditions is extremely
unusual in Santa Barbara and is potentially lethal to some

organisms. As a result of the rains we suspect that large quantities of DDT were flushed into the sea--the adverse biological effects of this insecticide are well documented. Thus a decision regarding the causes of mortality or redistribution of marine organisms during this period is not always straightforward. The reluctance of biologists to assign oil contamination as the cause of mortality or biological change is resented by the public who feel intuitively that all mortality or change should be attributed to oil.

Let us now examine some of the biological effects which have been observed since January 28, 1969. The most visible and immediate effect of the oil spill was the contamination of marine birds with oil. Hundreds, perhaps thousands of oil-soaked dead birds were reported from the beaches. How many additional birds died but did not reach shore remains unknown. At the time of this discussion Mrs. Gillian Sanders informed me that approximately 1500 living oil-contaminated birds were brought into the Santa Barbara bird rescue center for treatment. These consisted mostly of diving birds (grebes, cormorants, scoters, loons, pelicans and mergansers); only a very few gulls were brought in. The survival of these, even with treatment, was very low, amounting to roughly 10-12% according to Mrs. Sanders. Thus there has been an appreciable mortality of sea birds which can be attributed directly to oil. A number of large mammals and their young (seals, one porpoise, sea elephants), have been reported dead in the Santa Barbara area (including the Channel Islands). In most cases the causes of death could not be positively established although it seems possible that oil contamination may be a factor.

The effects of the oil spill on the commercial and sport fishery has been hotly debated in the press. At the present moment I don't think sufficient data have appeared in print to permit a sound evaluation of the effects of the oil spill upon the local fishery.

A survey of the distribution and abundance of fish eggs and larvae in the Santa Barbara Channel made by the Bureau

of Commercial Fisheries (La Jolla Laboratory) in February
showed a normal situation. Similarly, sampling of the
larger zooplankton by Dr. A. Ebeling and associates of UCSB
with financial assistance from the FWPCA showed an essentially normal biota, qualitatively and quantitatively.

A single observation of phytoplankton biomass made in
the vicinity of Platform A by the BCF was much less than
had been observed at a station on previous occasions located
some ten miles away from the Platform. My own observations
of net phytoplankton in Goleta Bay some 6 miles west of
Platform A in March, April, and May of this year showed
similar species composition and probably abundance to that
observed one year previously. Thus in Goleta Bay, which
was not as heavily contaminated as many local areas, the
effects of oil pollution upon the phytoplankton cannot be
detected with the methods employed.

On the beaches and in the intertidal the biological
effects of oil can be readily observed. On many rocky
surfaces the entire plant and animal communities have been
killed by a layer of encrusting oil which is often 1-2
centimeters thick. We have not observed any recolonization
of either animals or plants on these oil covered surfaces.
On other rocky surfaces a pronounced mortality of some of
the high-living barnacles of the genus Cthalamus have been
observed. Oil adheres readily to the local eel grass (a
higher plant) and a species of red alga (Endocladia sp.)
resulting in the death of these species. Although oil is
trapped in massive quantities in the kelp (Macrocystis)
beds it does not seem to have caused appreciable damage to
healthy plants. Otherwise intertidal organisms do not at
the present time exhibit severe damage due to pollution.
Some effects may take much longer to become visible than
the $3\frac{1}{2}$ months since January 28, 1969. Similarly, there
may be effects upon reproduction, migratory behavior, and
life cycle phenomena which have been unnoticed or undetectable
in the initial survey work. Continuing study is required
for some time in the future to evaluate the full impact of
the oil spill upon the biology of the Santa Barbara area.

As far as I can learn no study has been made of physical
and chemical effects of the oil films on the sea surface
created by the Union blowout. Floating oil affects the
coupling between the winds and the sea surface, prevents
evaporation, inhibits gaseous exchange between the atmosphere and sea, changes the albedo of the sea surface, and
modifies the quality and amount of solar radiation penetrating
the sea surface. All of these effects require study and some
of them may be of considerable importance.

The citizens of the Santa Barbara area are outraged
at the damage caused by the oil spill and dismayed at the
lack of a permanent Federal ban on further drilling in the
Channel. Several citizen's committees have been formed
such as GOO (Get Oil Out), COPE (Committee on a Pure Environment) to mention two of the more active groups to exert
pressure in the political arena. Officials in city, county
and state government, as well as local congressmen and
senators have been active in trying to change governmental
policies to prevent recurrence of the Santa Barbara disaster.
The apparent close relationship between the U. S. Geological
Survey and oil companies has come under sharp criticism and
must be carefully and impartially examined. The value,
effect, and purpose of oil subsidies such as the oil
depletion allowance have been carefully examined by econo-
mists and are now for the first time becoming important
public issues. Thus, the oil spill in Santa Barbara is not
only a serious local problem, but has attracted and continues
to attract national attention. Hopefully, enlightened public
policy may result which will benefit all citizens of our
country.

THE ROLE OF CHEMICAL DISPERSANTS IN OIL CLEANUP

Gerard P. Canevari

Esso Research and Engineering Company

P.O. Box 101, Florham Park, New Jersey

In order to place the role of a chemical dispersant for oil spills in the proper perspective, it should be pointed out that there are really three distinct areas of technical interest.

Firstly, during the normal operation of tankers, any water effluent is essentially oil-free. One method of achieving this is the load-on-top (LOT) procedure wherein oily ballast water is retained aboard and the incoming cargo loaded on top of it. This system was adopted by major carriers of petroleum several years ago. In addition, in a further effort to improve water quality, development work is in progress on oily-water separators, chemical flocculant to clarify tank washings and other methods of treatment suitable for shipboard use. In extraordinary cases where a tanker cargo compartment may be in danger of breaking up, additional development work is underway on gelling compounds to solidify the oil cargo in situ and thereby prevent its release. Figure 1 illustrates this technique and shows samples of gelled crudes made under both laboratory and field conditions.

Secondly, if there is an inadvertent accidental spill, it is generally agreed that the recommended procedure is to contain the oil and physically remove it with or without the aid of absorbents. Figure 2 depicts a field demonstration of polyurethane foam to absorb spilled oil in a quiescent area.[1]

Figure 1

GELLATION OF CRUDE OIL TO PREVENT
LEAKAGE FROM TANKER

Experimental Samples Of Laboratory Gelled Crudes

Surface Of Oil Gelled In 1000 Gal. Tank

(Note Block Resting On Surface)

Figure 2
FIELD TEST OF POLYURETHANE FOAM

NOTES:

    (1) The above use of polyurethane foam for absorbing oil spills has been developed in the U.K., at the Fawley Refinery of Esso Petroleum Co., Ltd., in conjunction with J. Bibby and Sons, Ltd.

    (2) Polyurethane foam has the following advantages for this application:

- The foam absorbs over 90% of its own volume of oil or 100 x its own weight.

- It is relatively inexpensive.

- It is easily generated by mixing two liquids to produce a hundred-fold expansion in one minute.

- The rate of water absorption is low.

- The oil-soaked foam does not release oil unless squeezed.

Figure 3

SPREADING FORCE FOR OIL ON WATER

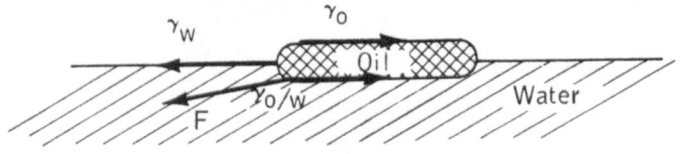

F, spreading force for oil $= \gamma_w - \gamma_o - \gamma_{o/w}$

Measured values for Kuwait Crude Oil on Sea Water

$$F = 61 - 28 - 22$$
$$= +11 \text{ ergs/cm}^2$$

∴ Oil will spread

OIL SPILL APPEARANCE

Film Thickness Inches x 10^{-6}	Appearance of Oil Films	Approximate Gallons/Sq. Mile
1.5	Barely Visible	25
3.0	Silver Sheen	50
6.0	First Trace of Color	100
12.0	Bright Bands of Color	200
80.0+	Much Darker Colors	1330+

Thirdly, if containment and recovery are impossible because of weather, sea conditions, etc., then consideration should be given to treating the oil slick at sea before it can cause damage. The objective is to remove the intact slick from the surface of the water either by burning, sinking or dispersing.

This paper will discuss the role of dispersants and will cover:

- The behavior of an oil spill on water

- The detrimental effects of the spill and basis for dispersing

- The mechanism and results of dispersing

- Tests of relative effectiveness of dispersants

- Toxicity studies of dispersants

- Some practical aspects of field application

The Behavior of an Oil Spill on Water and its Detrimental Effects

When a volume of oil is spilled onto water, it has a driving force to film out or spread -- in essence, a spreading pressure. It approaches an equilibrium thickness that is determined by a balance of surface forces illustrated by Figure 3 and expressed as follows:

$$F = \gamma_w - \gamma_o - \gamma_{o/w} \qquad (1)$$

where:

F = Spreading force or coefficient of oil on water, ergs/cm^2

γ_w = Surface tension of water phase, dynes/cm or ergs/cm^2

γ_o = Surface tension of oil phase, dynes/cm or ergs/cm^2

$\gamma_{o/w}$ = Interfacial tension of oil/water, dynes/cm or ergs/cm^2

It can be noted from measured values of Kuwait Crude and
sea water that the oil has a positive spreading coefficient
and will spread on the water phase. Similar data on additional crude oils have been published by S. A. Berridge
et al.[2]

The degree and rate that the oil spreads are based on
several factors that are beyond the scope of this paper.
However, it is interesting to note that the visual appearance of the oil film is an indication of its thickness. It
is difficult to estimate film thickness after the onset of
the dark colors since there is no change in appearance with
increasing oil slick thickness at this stage.

The resultant intact cohesive film of oil is detrimental
for reasons such as the following:

1. Marine fowl are particularly vulnerable to oil spills.
   The estimated minimum number of birds killed as a result of the Torrey Canyon grounding was 10,000. In
   this regard, it is interesting to note that extensive
   knowledge concerning the treatment of fouled birds
   was established by this unfortunate incident. However, despite such data as that published by Dr. Göran
   Odham, Göteborg, Univ., Sweden[4] that pinpointed the
   importance of featherwax and effective agents to re-
   habilitate birds, there are still instances of experi-
   mentation with various off-the-shelf toxic chemicals
   for this purpose.

2. Shore property and beaches can be extensively contaminated.

3. Slow moving crustaceans and intertidal marine life
   can be physically fouled by heavy coats of oil. There
   is also a particularly deleterious effect on the gill
   filaments of fish preventing the exchange of gases and
   resulting in anoxia.

4. There is evidence that the oil film forms a barrier to
   the transfer of oxygen into the water to support marine
   life, particularly planktonic species that reside less
   than 2-3 feet below the surface.

The Mechanism of Dispersing: A Properly
Selected Surfactant Promotes Fine Droplet Formation

The application of a properly selected surface-active
agent (surfactant) promotes further thinning of the oil
film since the oil/water interfacial tension ($\gamma_{o/w}$) is
reduced. More important, as depicted in Figure 4, the
chemical promotes fine oil droplet formation upon the application of mixing energy. It is important to emphasize
this latter point since lack of mixing has been one of the
principal reasons for reported instances of ineffective
dispersant application.

Since the surfactant consists of water compatible
(hydrophilic) and oil compatible (lipophilic) portions as
schematically shown in Figure 4, it arranges itself at the
oil-water interface in preference to either bulk phases.
By reducing oil/water interfacial tension ($\gamma_{o/w}$) in this
manner, the formation of additional interfacial areas
through fine droplet formation is promoted since:

$$W_K = A_{o/w} \; \gamma_{o/w} \qquad (2)$$

where:

W_K is mixing energy, ergs

$A_{o/w}$ is interfacial area, cm^2

$\gamma_{o/w}$ is interfacial tension, dynes/cm

There are many surfactants that will aid in the formation of fine droplets. An additional and more subtle requirement for an effective dispersant is the prevention of
coalescence of the oil droplets after they are formed. In
essence, the hydrophilic portion of the agent must parry oil
droplet collisions physically. Although the dispersed droplets may ultimately rise and concentrate at the surface in
a sample bottle, they should not coalesce to restore an intact oil film.

This property of an effective dispersant also prevents
the dispersed oil droplets from adhering to any surface,
e.g., beach sand, piers, bird's feathers, etc. An illustra-

Figure 4

EFFECT OF SURFACE ACTIVE AGENT (DISPERSANT)
ON OIL FILM

Surface active agent decreases $\gamma_{o/w}$ and increases spreading force, F

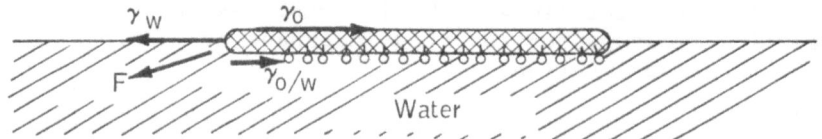

UPON APPLICATION OF MIXING ENERGY,
FINE OIL-IN-WATER DISPERSION IS FORMED

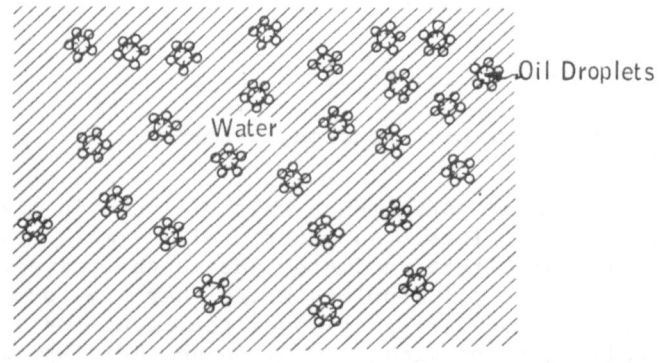

SCHEMATIC OF SURFACE ACTIVE AGENT

Water Compatible Portion
Oil Compatible Portion

Figure 5

CHEMICAL DISPERSANT PREVENTS OIL
FROM ADHERING TO SURFACE

Oil Film On Water With No Dispersant   Chemically Dispersed Oil

(Oil coats surface)   (Note oil does not adhere)

Figure 6

DISPERSANT'S EFFICIENCY MEASURED BY TURBIDITY

Laboratory Colorimeter to measure turbidity.

Note several cells of oily water dispersion on console.

Light Source   Oily Water Dispersion   Transmitted Light Measurement

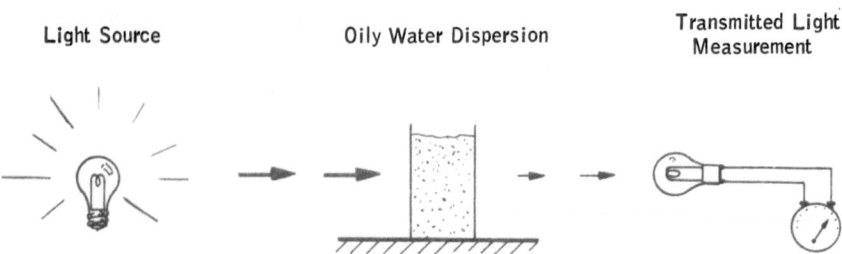

SCHEMATIC DIAGRAM OF TURBIDITY MEASUREMENT
BY LIGHT TRANSMISSION

tion of this behavior can be seen in Figure 5. The chemically dispersed oil has no tendency to adhere to a man's finger after immersion in the beaker but, as expected, an oil film is readily deposited from the beaker of untreated oil.

Dispersing Accelerates
The Biodegradation Rate

In addition to eliminating the detrimental physical aspects of an intact oil film, the orders of magnitude increase in interfacial area that are effected by the dispersant greatly increases the rate of biodegradation of the oil. Thus, whatever happens to reduce oil in the sea -- mainly bacterial action -- is greatly enhanced.

As pointed out by Dr. Claude Zobell,[5] the oil is made available to a much larger population of marine microorganisms. In addition, the 5 to 10 micron diameter droplets are more readily used by the marine bacteria (usually about 2-4 micron size).

Dr. ZoBell[5] indicates that the rate at which microorganisms oxidize hydrocarbon is influenced largely by the dispersion or solubility of the hydrocarbon and by the water temperature. The hydrocarbon can be used by the microorganism only if there is contact of the hydrocarbon with water. Since most hydrocarbons are only poorly soluble in water, their utilization depends upon emulsification or other means of dispersion in water. In general, an average of one-third of the hydrocarbons may be converted into bacterial cells. The remaining two-thirds of the hydrocarbon is oxidized largely to CO_2 and H_2O.

The Relative Effectiveness of a Dispersant
Determined by Laboratory Turbidity Measurements

It is difficult to assess dispersant effectiveness other than by actual tests on a freely moving body of water. One approach that has been found useful in determining relative effectiveness, however, is shown in Figure 6. It consists of mixing a given amount of oil and sea water under controlled conditions. The oily water dispersion is then poured into colorimeter cells for measurement of turbidity

as determined by light transmission. The change in light
transmission versus time is a measure of the stability of
the dispersion.

As graphically illustrated by Figure 7, the more stable
dispersion will transmit less light and this indicates a
more effective dispersant. It can be seen that within ten
minutes, the crude oil and water mixture without any dispersant is essentially separated as indicated by the large
amount of light transmission. This dispersion is not stable
because of the large oil droplet diameter and the rapid
coalescence of these droplets.

An additional test method which can be a worthwhile
supplement to the above consists of mixing the chemical
dispersant, oil and water in a small vessel and then adding
this dispersion to a larger volume of sea water. Air is
then bubbled through the mixture. The dispersant's effectiveness is determined by the amount of time that elapses before
a slick becomes visible. For example, as shown in Figure 8,
without a dispersion a slick formed immediately. In general,
this test correlates with the previous test results on the
relative effectiveness of the various dispersants.

Toxicity is an Equally Important Consideration for Any Chemical Introduced into the Marine Environment

The toxicity of the chemical dispersant is a consideration that is as important as effectiveness. This has been
the area of greatest concern and controversy; in fact, there
has been an almost evil connotation developed about the word
"chemical" for this use.

There is some basis for this concern. It was high-
lighted by the Torrey Canyon spill and was well documented
by the report of the Marine Biological Laboratory of the U.K.
at Plymouth, England.[6] This concise work indicated that
in some areas, particularly in the intertidal zone, the
chemicals used were more harmful than the oil itself.

The very extensive literature search conducted by
Battelle[3] also indicated that the biological effect of all
detergents is similar. It cited that a number of investiga-
tors (13 references) agreed that concentration of 5 to 10

Figure 7

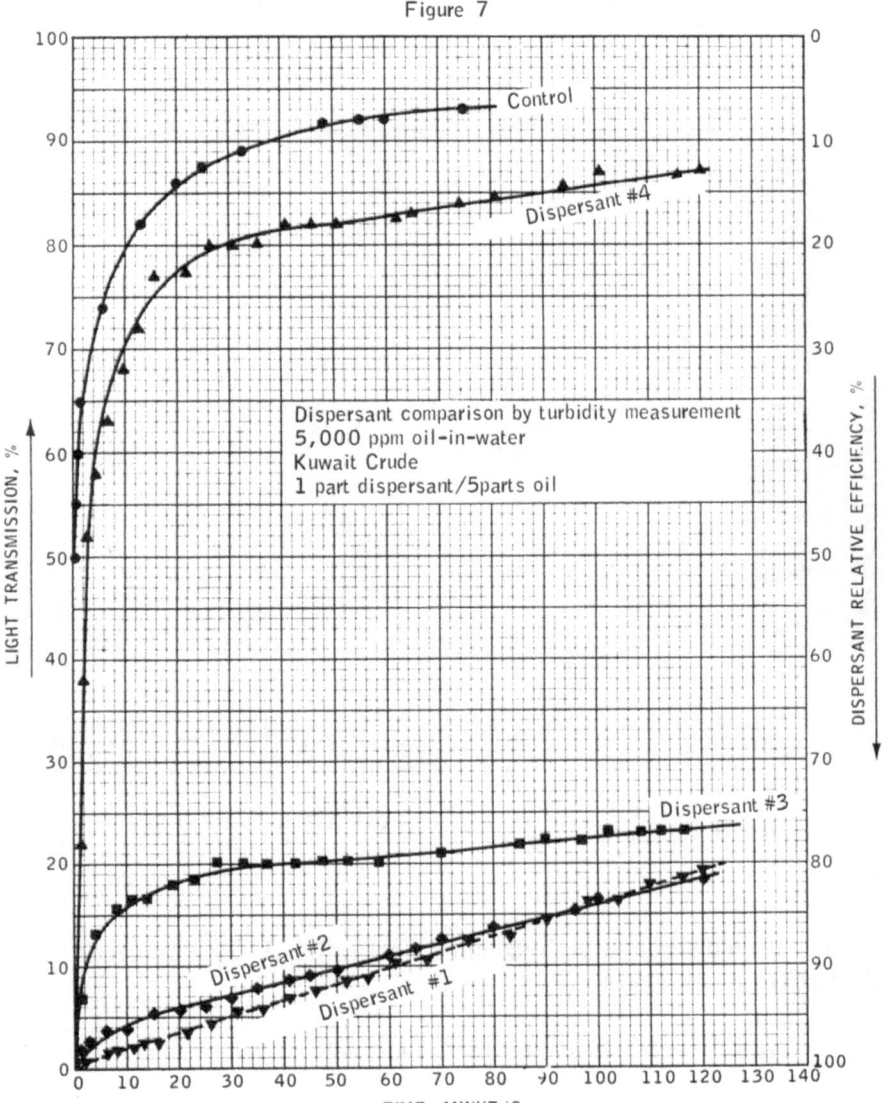

Gerard P. Canevari

Figure 8

LABORATORY METHOD DESIGNED
TO DETERMINE DISPERSANT'S EFFICIENCY

No Dispersant   Oil Slick Dispersant
                              Added

Air is pumped through both samples containing 1-1/2 cc crude oil in 3000 cc sea water.

Note stable dispersion formed by dispersant.

Figure 9

ONTARIO WATER RESOURCES COMMISSION
LAB. EVALUATION OF FIVE CHEMICAL DISPERSANTS

AUGUST, 1968

Chemical	TL$_{50}$ Concentration Fathead Minnow PPM
A	6.1
B*	9.5
C	5.3
D*	13.0
Corexit 7664	10,000 (No Toxic Effect Shown)

* These chemicals were also tested at the University of
Puerto Rico – March 1968.

Figure 10

INSTITUTE OF MARINE SCIENCES, MIAMI
EFFECT OF DISPERSANTS ON MARINE FISH (GAMBUSIA)

Control (Untreated Sea Water) = $\dfrac{3 \text{ Dead}}{20 \text{ Tested}}$						
	Concentration, PPM					
	1	15	30	100	1000	10,000
Corexit 7664	—	—	—	—	—	$\frac{2}{21}$
Product D	$\frac{2}{20}$	$\frac{21}{21}$	$\frac{21}{21}$	—	—	—
Product E	$\frac{6}{21}$	$\frac{12}{22}$	$\frac{21}{21}$	—	—	—
Product F	$\frac{9}{20}$	$\frac{16}{21}$	$\frac{20}{20}$	—	—	—
Product G	$\frac{4}{21}$	$\frac{6}{21}$	$\frac{16}{21}$	—	—	—

ppm or more will cause death on the basis of a study of a
wide variety of fish.

Further, the Department of Marine Sciences, University
of Puerto Rico, conducted a survey of the Ocean Eagle Spill
(March 1966) and the effect of the four detergents used in
the oil cleanup.[7]   The results of the rather short duration bioassays showed that, at concentrations greater than
1 ppm, these detergents were toxic to some littoral species.
It was therefore recommended that the use of detergents be
discontinued at Puerto Rico.

Although, as cited above, the specific agents tested
did exhibit a high level of toxicity, this property is not
an inherent characteristic of surfactants. To illustrate
this, the results of toxicity tests conducted by the Ontario
Resources Commission are presented in Figure 9. It can be
seen that there is a major difference in the 96 hours TL_{50}
concentration (wherein there is a 50% mortality of the
animals). For example, COREXIT 7664 contains a non-ionic
surfactant and, while being an effective oil dispersant,
did not cause any harm to the test fish at concentrations of
10,000 ppm. Clearly then, the position that all chemical
detergents and dispersants are, in themselves, highly toxic
is incorrect.

COREXIT 7664 Oil Dispersant was developed by Esso
Research and Engineering Company as part of an overall R&D
Program in the field of Marine Conservation. Therefore,
this particular chemical is cited since an abundance of in-
house data on it is available.

To establish that this was not an isolated data point
or unique to the specific type of marine life tested (Fat-
heat Minnow), extensive toxicological studies were conducted
at the Institute of Marine Sciences, Miami. Representative
data is depicted in Figure 10 and substantiates the previously cited investigations at Ontario wherein COREXIT 7664
and Product D were tested.

A further detailed investigation into this area was conducted by the Marine Biological Laboratory, Plymouth, England
by Dr. Molly Spooner[8]. Some pertinent passages from the
abstract of this report follow:

"The relative toxicities of five kinds of dispersants
have been compared. COREXIT 7664, though not completely non-toxic, is far less toxic than any of the
other kinds investigated.

Planktonic species and a number of shore types have
been tested with these dispersants.

Tests with barnacle larvae showed that COREXIT 7664
is much less toxic than crude oil or than fresh
bunker oil. When dispersion into small droplets
takes place, there is an increase in the effective
toxicity of the oil when held in a confined vessel
but this phase is short lived, and could scarcely
occur under conditions of practical application.

Work on mussels showed that a state of extremely
low toxicity is reached within a few hours in
experimental jars under aeration. Mussels, stimulated to spawn by the somewhat unpleasant conditions, gave rise to both eggs, sperm and developing
larvae which survived for up to 48 hours in soupy
suspensions of various oils and COREXIT 7664."

It is interesting to note the experimental probe into the
effect of the dispersed oil on the marine environment.
Further work is continuing in this latter area to insure
that all possible effects of chemically dispersed oil are
completely understood and quantified to the maximum possible extent.

Some Practical Aspects
Of Field Application

It is important to emphasize that oil slick dispersants
are no panacaea. They have a role in oil spill cleanup and
specific limits of effective application within that role.
The following are pertinent considerations:

1. Water base or petroleum base dispersant system -
   Dispersant systems may utilize either a water solvent with a predominately water-compatible surfactant
   or a petroleum base solvent with the converse. This
   difference does influence the method of application.

   For example, an efficient and probably the most common method of dispersant application is by the eductor
   method. This equipment is designed for connection to
   fixed fire fighting systems aboard ships, boats or
   docks. A metering valve in the eductor permits 1 to
   6% of the dispersant to be educted into the water
   system. The water jet is an effective vehicle or
   carrier for the dispersant and permits good coverage
   in treating the oil slick. The high velocity fire
   hose may then be used to provide the required mixing
   energy. Figure 11 shows an application of a water
   base dispersant onto an oil slick wherein a 2%
   solution of COREXIT 7664 is educted into the fire-
   main system and then applied by means of a spray
   boom. The boat's bow wake and propeller provide
   the needed mixing energy.

   This application procedure, however, is somewhat
   incompatible with petroleum solvent systems, even
   though the latter type may be as efficient an oil-
   in-water dispersant as a specific water base system.
   This is because an oil solvent-in-water dispersion
   is formed as soon as the oil solvent base dispersant
   is educted into the fire hose. This accounts for
   the milky white appearance of the water after such
   application. In this state, it is difficult for the
   surfactant to transfer from its thermodynamically
   stable location at the oil solvent-water interface
   to the oil spill-sea water interface. Therefore,
   for a petroleum base dispersant system, neat application of the chemical directly onto the oil slick is
   a more effective application method.

2. Application of mixing energy - Mixing energy is required to form the dispersion after application of
   the dispersant. One of the most effective methods
   of application is by means of high pressure spray
   booms mounted on boats. As mentioned, the agitation
   from the wake of the boat is sufficient to effect
   dispersion when seas are calm. Dispersants can and

Figure 11

APPLICATION OF CHEMICAL DISPERSANT TO OIL SLICK

Figure 12
EFFECT OF OIL VISCOSITY ON STABILITY OF DISPERSION

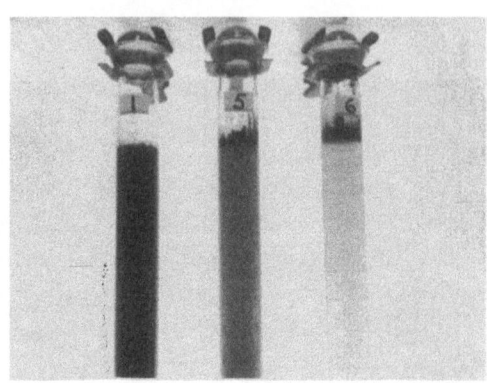

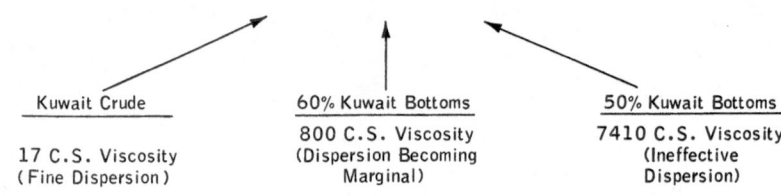

Kuwait Crude

17 C.S. Viscosity
(Fine Dispersion)

60% Kuwait Bottoms
800 C.S. Viscosity
(Dispersion Becoming
Marginal)

50% Kuwait Bottoms
7410 C.S. Viscosity
(Ineffective
Dispersion)

Note: 1 - All samples treated with $\dfrac{1 \text{ Part Dispersant (Corexit 7664)}}{10 \text{ Parts Oil}}$ and given

similar agitation.

have been applied effectively by spraying from aircraft provided there is sufficient agitation from
boats or from wave action. Aircraft and boats can
work in concert. Small spills in harbor areas can
be most conveniently handled by application from
high pressure fire hoses.

The prompt application of mixing energy is particularly important for a water base system. In essence,
small oil droplets must be produced while the immediate water environment is surfactant-rich. Surfactant
is then available to absorb at the newly created oil
droplet-water boundary to prevent coalescence.

3. <u>Viscosity of the oil spill</u> - Assuming that proper
   application procedures have been maintained, the
   principle factor affecting dispersant effectiveness
   is the viscosity of the oil spill. In this regard,
   in order to evaluate the effect of the viscosity
   parameter on dispersant performance (specifically
   COREXIT 7664), the following fractions of Kuwait
   Crude were distilled:

   - 90% Kuwait bottoms cut ------   35 cs
   - 80% Kuwait bottoms cut ------   88 cs
   - 70% Kuwait bottoms cut ------   260 cs
   - 65% Kuwait bottoms cut ------   800 cs
   - 50% Kuwait bottoms cut ------ 7410 cs

Based on the previously described lab tests, it was
determined, for this particular water base dispersant, that its effectiveness became marginal at
approximately 1000 cs; in essence, unless large
amounts of mixing energy are applied, some globules
of oil, which readily come to the surface, are formed
rather than homogenous fine droplets at this level
of viscosity.)

This is illustrated in Figure 12 where samples of
Kuwait Crude (17 cs viscosity), 60% Kuwait Bottoms
(800 cs viscosity) and 50% Kuwait* Bottoms (7410
cs viscosity) have been dispersed in sea water by

<u>NOTE</u>: * 50% Kuwait Bottoms (7410 cs viscosity) represents
       a tar-like, almost immobile fluid.

the addition of COREXIT 7664 (1 part COREXIT/10
parts oil). After a uniform amount of agitation,
a gradation in the stability of the dispersions
can readily be seen.

Although data has been presented for one particular
dispersant, this consideration is applicable to
dispersants in general. (There may be some slight
variations in the specific viscosity levels.)
However, it should be emphasized that an oil viscosity limit of 1000 cs does not represent a
critical limitation since over 90% of tanker
cargoes carried world-wide have viscosities well
below 1000 cs.

Conclusion and Future Effort

In summary, it must be appreciated that we are not
dealing with a comparison of a clean water environment
versus a marine environment with chemically dispersed oil
droplets. Rather, the choice is between an intact, cohesive oil slick versus treatment such as dispersion as
illustrated in Figure 11. One must objectively consider
all available data from the many ongoing studies in this
area to select the least objectionable situation.

As mentioned, investigations are continuing at many
locations on the biological effect of the oil-dispersant
mixtures. The data on the effect of the dispersant alone
is still very relevant since experience has shown that there
is a tendency to "overtreat" and apply excessive amounts of
chemical during an oil spill emergency. Hence there is
always the probability that some dispersant will escape
into the marine environment without associating with the
oil spill. It is therefore most important that the chemical,
in itself, is not any more toxic than the oil itself as a
minimum requirement.

REFERENCES

1. Mayo, F., "Dealing with Pollution on Water and Shores." Institute of Petroleum Meeting, Brighton, England (1960).

2. Berridge, S.A., Dean, R. A., Fallows, R. G., and Fish, A., "The Properties of Persistent Oils at Sea." Institute of Petroleum Symposium (1968).

3. Battelle Memorial Institute, "Oil Spillage Study." U.S. Department of Commerce Report AD 666289 (1967).

4. Odham, Göran, "Oiled Water Birds - New Possibilities for Rehabilitation," Department of Medical Chemistry, Goteborg University, Goteborg, Sweden (1968).

5. ZoBell, C., "The Occurrence, Effects and Fate of Oil Polluting the Sea," J. Air Water Poll., Pergamon Press (1963).

6. Smith, J.E., "Torrey Canyon-Pollution and Marine," Cambridge University Press (1968).

7. Cerame-Vivar, M.J., "The Ocean Eagle Spill," Department of Marine Science, University of Puerto Rico (1968).

8. Spooner, M. F., "Preliminary Work on Comparative Toxicities of Some Oil Spill Dispersants and a Few Tests with Oils and COREXIT," Marine Biological Laboratory, Plymouth (1968).

THE SPREAD OF OIL SLICKS ON A CALM SEA

James A. Fay

Department of Mechanical Engineering
Massachusetts Institute of Technology
Cambridge, Massachusetts

I. PHYSICAL CONSIDERATIONS

It is a common observation that oil, when spilled on water, tends to spread outward on the water surface in the form of a thin continuous layer. In those instances where this layer is as thin as a wave length of visible light, an iridescent color of the film, caused by light interference, is observed. This tendency to spread is the result of two physical forces: the force of gravity which causes the lighter oil to seek a constant level by spreading horizontally, just as it would on a plane horizontal solid surface, and the surface tension force of pure water, which is usually greater than that of the oil film floating on water. While the oil layer could spread while still remaining intact until it had formed a monomolecular layer, spreading usually stops when the layer is much thicker than this, most likely because of a change in the surface tension properties of the oil.

It might appear to be a paradox that the force of gravity, which acts downward, should cause a layer of oil to spread sideways. The horizontal motion is actually caused by outward pressure forces in the oil, but these pressure forces are themselves a result of the vertical gravitational force. It is perhaps easier to understand this spreading tendency if we note that a floating layer of oil (as indeed for any floating body of uniform density) has a more elevated center of gravity, and hence greater potential

53

energy, than the fluid it displaces. Thus the gradual
spread of an oil slick is accompanied by a loss of potential
energy in the earth's gravitational field.

A similar loss of surface energy occurs as an oil slick
spreads. For a given increase in slick area, the energy of
the air-water interface (or surface tension) is lost while
those of an equal area of air-oil and oil-water interfaces
are gained. For an oil which "wets" the water, and hence
spreads, this results in a net loss of surface energy.

To conserve energy in the spreading process, the loss
of gravitational energy and surface energy must be balanced
by a gain in other forms of energy. There will be a gain in
the kinetic energy of the moving oil and water and an increase in internal (heat) energy generated by viscous forces
in both oil and water. In terms of forces, therefore, we
expect gravity and surface tension to increase the spread
while inertia and viscous forces retard it.

In the open sea, this spreading tendency is aided by
water surface motions induced by waves, wind and tidal currents. These surface motions are characterized by random-
ness in strength and direction, so that different elements
of a slick are moved about relative to each other and to the
center of mass of the slick, even when there is a gradual
drift of the slick as a whole due to wind, wave or current.
Therefore there can be a dispersal of a slick caused by such
random motions, which is quite analogous (in two dimensions)
to the three dimensional dispersion of a puff of smoke
emitted into the atmosphere on a windy day. This will certainly be the only cause of spread if the slick has been
broken into small independent pieces, each of which has
reached a stable size. The spread resulting from this random motion of the sea surface is very difficult to estimate,
but appears to be smaller, under conditions for which
observations have been made, than that caused by tension and
gravity forces.

Most oils spilled are mixtures of components having
varying vapor pressure and solubility in water. When
initially spilled, the fractions of lighter molecular weight,
being more volatile and soluble, are preferentially leached
out, leaving a residue which is denser and more viscous.
Because of the low molecular diffusivity of liquids, this is
a slow process. Thus it can be expected that the bulk

properties which are important to the spreading, such as
density, viscosity and surface and interfacial tensions,
will change slowly with time while the slick spreads.

II.   THE RATE OF SPREAD OF A FINITE QUANTITY OF OIL ON STILL WATER

For the present purposes we shall be satisfied to
estimate the order of magnitude of the rate of spread of an
oil slick on the surface of still water, i.e., water which
is free of motions induced by wind, wave and tidal currents.
We shall assume that, at zero time t, a volume V of oil is
dumped on the water, and that it subsequently spreads outward in a slick whose diameter ℓ and thickness h change with
time t.   In order of magnitude, to conserve the volume of
oil,

$$V = \ell^2 h \tag{1}$$

As it spreads outwards, the moving layer of oil drags
with it a thin layer of water, because the oil cannot slip
across the water surface.   (The water is thus not entirely
"still" during the spreading process, for it moves upward to
replace the spreading oil as well as outward near the oil-
water interface due to the viscous drag of the oil film.)
The thickness δ of the uppermost layer of water so set into
motion by viscous forces has the magnitude

$$\delta = \{\nu t\}^{1/2} \tag{2}$$

in which ν is the kinematic viscosity of the water.   This is
the thickness of the boundary layer at the edge of a fluid
in which the viscous force can accelerate the fluid up to
the speed of a moving boundary (in this case, the outwardly
moving oil film).

Based on this model, we shall next estimate the order
of magnitude of the forces (per unit volume of oil) which
tend to accelerate or retard the spread.   To this end, we
first determine the surface tension force per unit volume of
oil by dividing the net surface tension σ by the cross-
sectional area of the slick, ℓh, and then use Eq. (1) to find

Surface tension: $\sigma/\ell h = \sigma\ell/V \tag{3}$

The other spreading force, gravity, produces a horizontal

force per unit volume (pressure gradient) of $\Delta\rho gh/\ell$, or using Eq. (1),

Gravity: $\Delta\rho gV/\ell^3$ (4)

in which $\Delta\rho$ is the difference in mass density between water and oil and g is the gravitational acceleration. The inertia force, which retards the flow, is the product of mass density ρ and acceleration ℓ/t^2:

Inertia: $\rho\ell/t^2$ (5)

The viscous force per unit volume is the viscous stress $\rho\nu(\ell/t)/\delta$, or product of absolute viscosity $\rho\nu$ and velocity gradient $(\ell/t)/\delta$ in the water, divided by the film thickness h:

Viscous: $\rho\nu(\ell/t)/\delta h = \rho\nu^{1/2}\ell^3/Vt^{3/2}$ (6)

in which Eqs.* (1) and (2) were used to derive the expression on the right*.

    Now consider the two forces which tend to spread out the slick, surface tension (Eq. (3)) and gravity (Eq. (4)). Since ℓ increases monotonically as time passes, gravity will be the dominant spreading force at small times while surface tension will eventually dominate at long times. By setting the two forces equal to each other, and using Eq. (1), we may find the critical thickness h_c below which the spread of an oil slick is always dominated by surface tension:

$$h_c = (\sigma/\Delta\rho g)^{1/2}$$ (7)**

We may thus conclude that, early in the spreading process before the slick has thinned to h_c, gravity dominates while

_____

*   Because this is only an order of magnitude estimate, we don't distinguish between the densities of water and oil (calling both ρ) except when their difference $\Delta\rho$ appears directly, as in Eq. (4).

** For typical values of σ and $\Delta\rho$ of 30 dynes/cm and 0.05 gm/cm^3 respectively, h_c is 0.8 cm.

surface tension causes spreading for films much thinner than h_c.

Now consider the retarding forces of inertia (Eq. (5)) and viscosity (Eq. (6)). The ratio of viscous to inertia force varies as $\ell^2 t^{1/2}$ and must therefore become very small near the beginning of the spread. Thus early in the spreading process inertia forces dominate, while ultimately viscous forces will produce the major retarding effect. By reference to Eqs. (1) and (2), it can be seen that these two forces become equal when the film thickness h equals the thickness δ of the viscous layer in the water. This transition from inertially retarded to viscously retarded spread does not occur for a fixed thickness, nor at a given time, but varies with the size of the spill.

The transition from gravity to surface tension spread, and inertial to viscous retardation of spread, would occur simultaneously when $\delta = h_c$ or $t = h_c^2/\nu$ *. This coincidence is unlikely to be achieved outside of the laboratory. In most large scale oil spills in the open, by the time the slick has thinned to the thickness h_c, δ is much greater than h_c, so that the viscous retarding force sets in before the surface tension spreading force becomes important. The history of the spread then passes through three phases:

(i)   the beginning phase in which only gravity and inertia forces are important,

(ii)   an intermediate phase in which gravity and viscous forces dominate and

(iii)   a final phase in which surface tension is balanced by viscous forces.

Gravity-inertia:	$\ell = (\Delta g V t^2)^{1/4}$	(8)
Gravity-viscous:	$\ell = (\Delta g V^2 t^{3/2}/\nu^{1/2})^{1/6}$	(9)
Surface tension-viscous	$\ell = (\sigma^2 t^3/\rho^2 \nu)^{1/4}$	(10)

---

* For $\nu = 10^{-2}$ cm^2/sec and $h_c = 0.8$ cm, $t = 60$ sec.

To illustrate the respective regimes of the different phases of spreading, we have plotted Eqs. (8) - (10) in Fig. 1 for the special case of a 10,000 ton spill of oil (about the size of the Torrey Canyon initial spill). The very slow spread of the slick, and the domination of viscous retardation after the first hour of spill life, are clearly illustrated. Of course, since these are only order of magnitude estimates, the transition point from one phase to another, as well as the exact size of the spill at a given time, are only approximately given by Eqs. (8) - (10).

For smaller spills, the third phase of spread (surface tension-viscous) becomes dominant earlier in the history of the spread. For practical purposes, it may be sufficient to use Eq. (10) describing the final phase for any time later than the first hour or two after initiating the spill.

The final phase of spreading (Eq. (10)) shows a growth which is independent of the volume V of the oil spill. This is a result of the fact that the thickness of the slick is no longer important in determining the major forces.

III. SPREAD OF A SLICK FROM A STEADY SOURCE IN A MOVING STREAM

With only slight modifications, it is possible to estimate the width ℓ of a slick spreading from a stationary source of volume flow rate $\dot{V}$ located in a stream of uniform speed u. This would be a suitable model for an oil well or grounded tanker leaking steadily into a sea with uniform tidal current. We may use the force estimates previously described provided we replace time t by the flow time x/u, where x is the downstream distance from the source, and the conservation of oil Eq. (1) by

$$\dot{V} = h\ell u \qquad\qquad (11)$$

With these changes the surface tension, gravity, inertia and viscous forces become:

Surface tension: $\sigma u / \dot{V}$ (12)

Gravity: $\Delta \rho g \dot{V} / \ell^2 u$ (13)

Inertia: $\rho \ell u^2 / x^2$ (14)

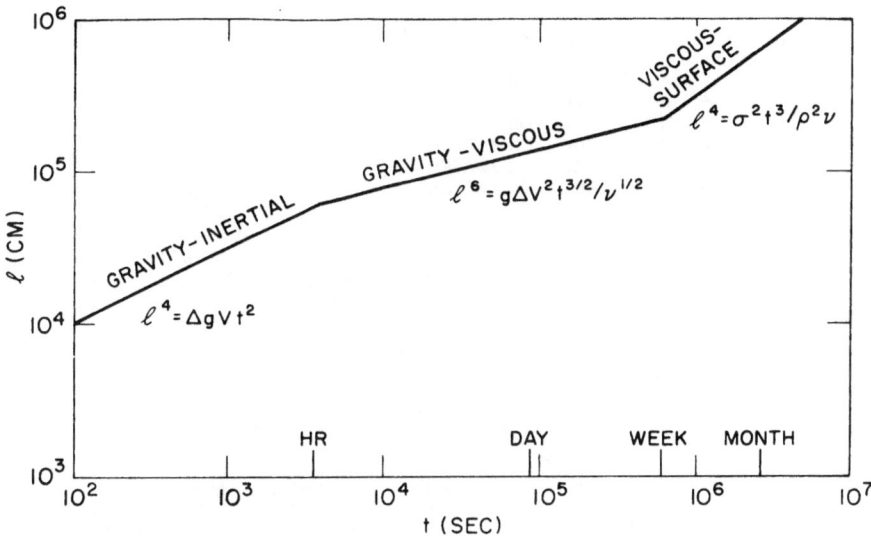

Figure 1. The size ℓ of an oil slick as a function of time t for a 10,000 ton spill.

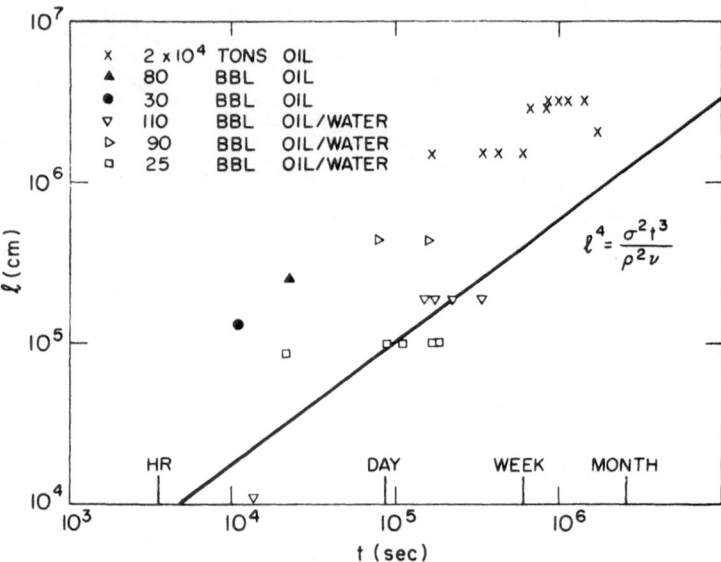

Figure 2. A comparison of measured oil slick size and the theoretical estimate for surface tension-viscous spreading.

Viscous:
$$\rho v^{1/2} \ell^2 u^{5/2} / \dot{v} x^{3/2} \tag{15}$$

The three spreading regimes are readily found to be:

Gravity-inertia: $\ell = (\Delta g \dot{v} x^2)^{1/3} / u$ $\qquad$ (16)

Gravity-viscous: $\ell = (\Delta g \dot{v}^2 x^{3/2} / v^{1/2} u^{7/2})^{1/4}$ $\qquad$ (17)

Surface tension-viscous: $\ell = (\sigma^2 x^3 / \rho^2 v u^3)^{1/4}$ $\qquad$ (18)

As for the case of instantaneous spill, the final phase of spreading is independent of the strength of the source. Other remarks previously made, as to the significance of h_c and δ, apply equally well to this case.

IV.   COMPARISON WITH FIELD OBSERVATIONS

There are a very limited number of reports of observations of the spread of oil slicks on the open sea. Smith[1] gives data on the spread of the slick from the Torrey Canyon, while Stroop[2] describes small scale spill tests conducted by the U.S. Navy in 1927. We have plotted these observations of slick size ℓ as a function of time t since initiating the spill in Fig. 2. Also shown in Fig. 2 is Eq. (10) for the final phase of spread.

The observations almost always show a rapid spread to a size which increases with volume of the spill, followed by a long period of no further growth. In no case were observations made of the earlier, growing phase.

We ascribe this behavior to a sudden reduction of the net surface tension σ at a time when evaporation and dissolving of the lighter components of the oil has occurred. For thinner slicks (smaller volumes) this evaporation occurs sooner and leads to the cessation of spreading at an earlier time (and hence smaller size ℓ). If this is so, the observed sizes should all lie below a line of the form given by Eq. (10). This is approximately true of the data shown in Fig. 2 if we choose

$$\ell = 10(\sigma^2 t^3 / \rho^2 v)^{1/4} \tag{19}$$

as an upper bound for these points.

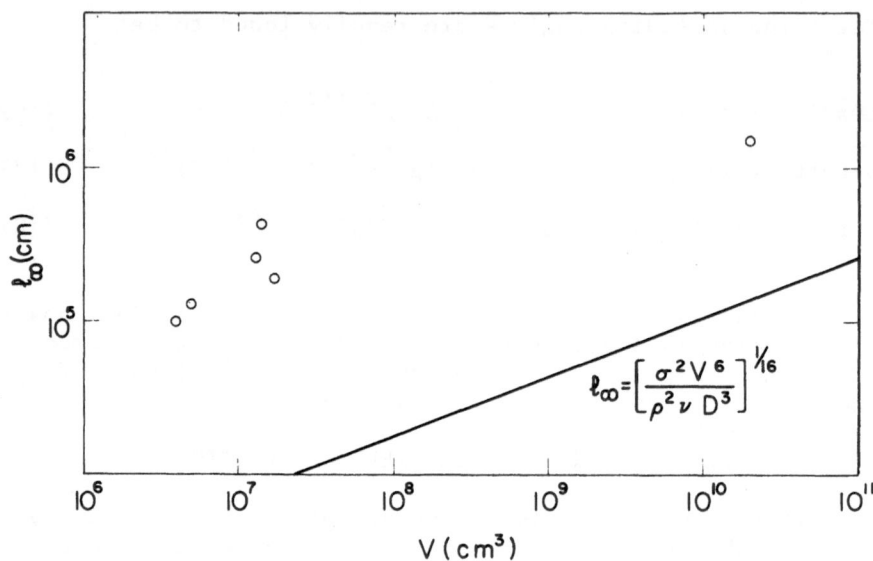

Figure 3. A comparison of measured and theoretical values for the final size ℓ_∞ of a slick as a function of the spill volume V.

We have tested this hypothesis by determining the thickness h for which molecular diffusion to the upper or lower surface of the oil film could have depleted components of the oil in the time t required to reach this thickness. Setting

$$h = \sqrt{Dt} \tag{20}$$

in which D is the molecular diffusivity of oil (about 10^{-5} cm^2/sec), and using Eqs. (10) and (1), we can solve for the scale ℓ_∞ at which Eq. (20) is satisfied:

$$\ell_\infty = (\sigma^2 V^6 / \rho^2 \nu D^3)^{1/16} \tag{21}$$

This theoretical relation is shown in Fig. 3, together with the measured size of slick taken from Fig. 2. Although the agreement is not good, the slow rate of increase of ℓ_∞ (and consequently h) with slick volume V is about as predicted by the theory.

REFERENCES

1. Smith, J. E., (ed.), 'Torrey Canyon' Pollution and Marine Life, Cambridge University Press (Cambridge, 1968).

2. Stroop, D. V., "Report on Oil Pollution Experiments – Behavior of Fuel Oil in the Surface of the Sea," pp. 41-49, Pollution of Navigable Waters, Bureau of Standards, Washington (1927).

CONTAINMENT AND COLLECTION DEVICES FOR OIL SLICKS

David P. Hoult

Department of Mechanical Engineering
Massachusetts Institute of Technology
Cambridge, Massachusetts

ABSTRACT

The main engineering features of physical and pneumatic
booms operating as oil containment devices in the open sea
are discussed. The oil containing capacity of a pneumatic
boom is limited by the power available to compress the air.
It is suggested that the containment capacity of physical
booms is limited by the effective depth of the boom. The
results of preliminary experiments are used to establish
these ideas. A brief review of the main types of collection
systems in use is given.

I. INTRODUCTION

Present containment devices for oil slicks consist of
two main types: physical booms or barriers, and pneumatic
booms. A typical physical boom might consist of a buoyant
cylinder, 1 ft in diameter, to which is attacked a fabric
curtain 1 1/2 ft deep, which is weighed down by a chain sewn
in its lower edge. In operation, it is supposed that combination of the buoyant cyclinder and the chain tend to hold
the curtain vertically in the sea, so as to produce a vertical
barrier to the spread of oil.

A pneumatic boom consists of a pipe, submerged below the
sea surface, which is supplied with compressed air. The air
is allowed to escape through small holes in the pipe, which

creates a large number of fine bubbles. As these bubbles
mix with the water, an air-water mixture is created whose
density is slightly less than that of the surrounding water.
Hence, the mixture rises due to buoyancy, creating a vertical
current. At the surface, this current splits into two surface currents, on opposite sides of the boom, and moving away
from it. It is this surface current which is used to buck
the tendency of the oil to spread.

Collection devices, which we shall discuss briefly, consist of two main types: a roller, and a towed or pushed
boom. The roller, as it moves over the surface of the sea,
collects the oil which sticks to it. As the roller rotates,
this oil is lifted out of the water, and scraped off into a
collector. Towed boom systems are based on the observation
that if a Vee shaped boom is pushed into an oil slick, oil
tends to pile up in the notch of the Vee, where it may be
collected.

It is a common observation of the Torrey Canyon and
Santa Barbara disasters that none of these devices worked
in the open sea. It is the thesis of this paper that these
devices fail because none are designed to operate in the
actual environment of the sea.

The main environmental conditions which must be taken
into account are the effects of winds, waves, and currents.
A typical wind velocity might be 40 ft/sec ($\sim$ 24 knots).
Such a wind generates a surface current of about 3% of the
wind speed[1], i.e. about 1.2 ft/sec. Oil on the calm sea
will be blown about by the wind with about this velocity.

In general, the wind will generate a wave field. If
the waves have a height 2a, then a particle on the surface
moves back and forth a distance 2a in one wave cycle. In
4 ft high waves, an oil slick sloshes back and forth 4 ft as
each wave goes by. The period of waves, generated by the
wind, may be estimated by realizing that the most intense
waves will be those whose phase velocity is equal to the wind
speed; for a 40 ft/sec wind, this gives waves of 10 sec
period.

Even without a wind, a particle in 4 ft high waves with
a 10 second period has a mean drift velocity[2] in addition
to the back and forth motion. This is a second order
effect of the wave field, and amounts to 1 ft/sec in this

case. With a wind present, the effects of the wind and wave
induced surface currents combine in some way to generate
surface currents in excess of 1 ft/sec.

Acting in combination with wind and waves are tidal
currents. Typical tidal currents have velocities from 1/2
to 2 1/2 knots or 3/4 ft/sec to 4 ft/sec.

These effects then tend to transport oil spilled on the
sea with velocities in excess of 1 ft/sec. It is the purpose of containment devices to combat the tendency of the
oil to spread due to wind, waves, currents, and those effects
of density difference and surface tension which tend to cause
oil to spread on still water.

II. THE PNEUMATIC BOOM

Consider a pipe, located at a depth d below the water
surface. The pipe is supplied with compressed air which
flows out of small holes drilled in the pipe. The bubbles
formed create a buoyant air and water mixture. (See Figure 1).

At some height z, above the pipe, the buoyant mixture
has a width b and rises with a velocity w. Let Δ be the
volume of bubbles in a unit volume of mixture at the height
z. If air flows out of the pipe at a rate V cu-ft/sec/foot
of pipe, since the volume of air is conserved,

$$V = wb\Delta \quad . \tag{1}$$

Now the buoyancy acting to accelerate upward the plume,
at a height z, is $\rho b\Delta g$, where ρ is the density of the water,
and g is the acceleration due to gravity. The balance between the buoyant force and the change in momentum in the
plume may be written as

$$\frac{d}{dz}(bw^2) = \rho b\Delta g \quad . \tag{2}$$

Using equation (1) to eliminate Δ then gives

$$\frac{d}{dz}(bw^2) = \frac{Vg}{w} \quad . \tag{3}$$

In still water, b $\sim$ z, and hence equation (3) yields an estimate[3] for w:

$$w \sim (Vg)^{1/3} \quad . \tag{4}$$

In rough water, b may depend on wave action, but it can be shown, as in equation (4), that w remains proportional to the cube root of the volume rate of flow.

In still water, when the buoyant plume reaches the surface, it divides into two horizontal jets. Let x be the horizontal distance from the pipe. (See Figure 2) The depth of the surface current is proportional to x. The magnitude of the surface current, U, is determined by the fact that the momentum in the surface current is conserved. Since the initial momentum is approximately $(Vg)^{2/3}d$, U is given by

$$U^2 x \sim (Vg)^{2/3}d \quad . \tag{5}$$

Clearly, this result only holds some distance from the pipe; the maximum value of U is

$$U_{max} \sim (Vg)^{1/3} \quad . \tag{6}$$

An experiment was performed in our laboratory[4] to check equation (5). Small wood chips were released at x = 0, above pneumatic boom, and the resulting position of the particle as a function of time was recorded. From equation (5) we expect that the particle position x(t) is given by

$$\frac{dx}{dt} = U \sim \sqrt{d/x} \, (Vg)^{1/3} \quad .$$

The result may be written as

$$\frac{(Vg)^{1/3}t}{d} \sim \left(\frac{x}{d}\right)^{3/2} \quad . \tag{7}$$

Figure 3 shows the experimental data which do in fact agree with equation (7).

Let us suppose that in a wave field, the surface current generated by a pneumatic boom is substantially the same as

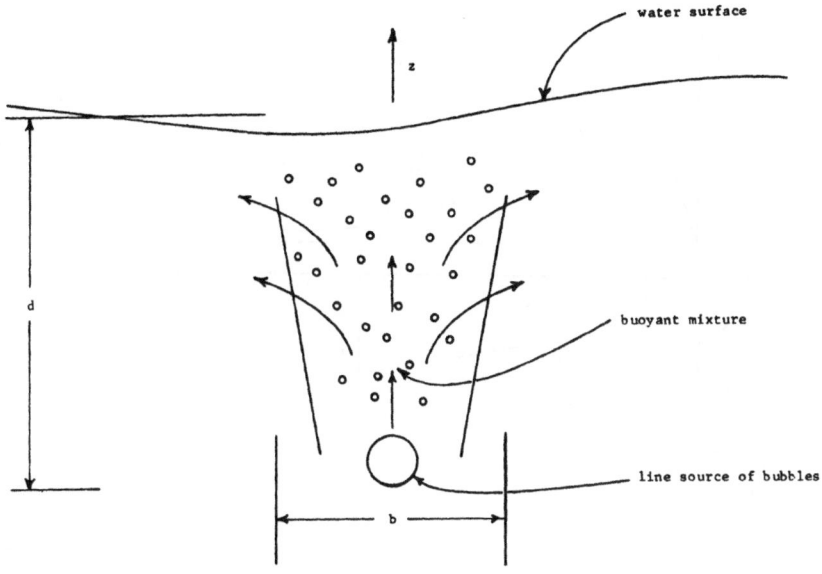

Figure 1. This sketch of a pneumatic boom shows the region where a mixture of air bubbles and water is formed above the source of compressed air. Since the mixture is slightly less dense than the water alone, there is a vertical current, created due to the action of buoyancy.

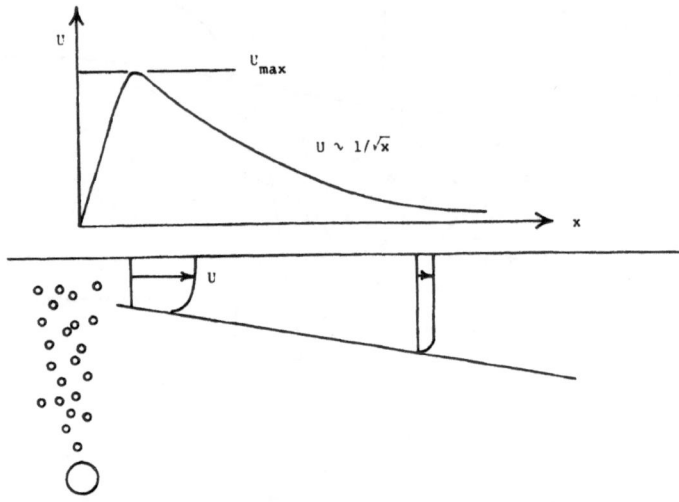

Figure 2. This sketch shows how the surface current generated by the line source of bubbles decays with distance
from the boom.

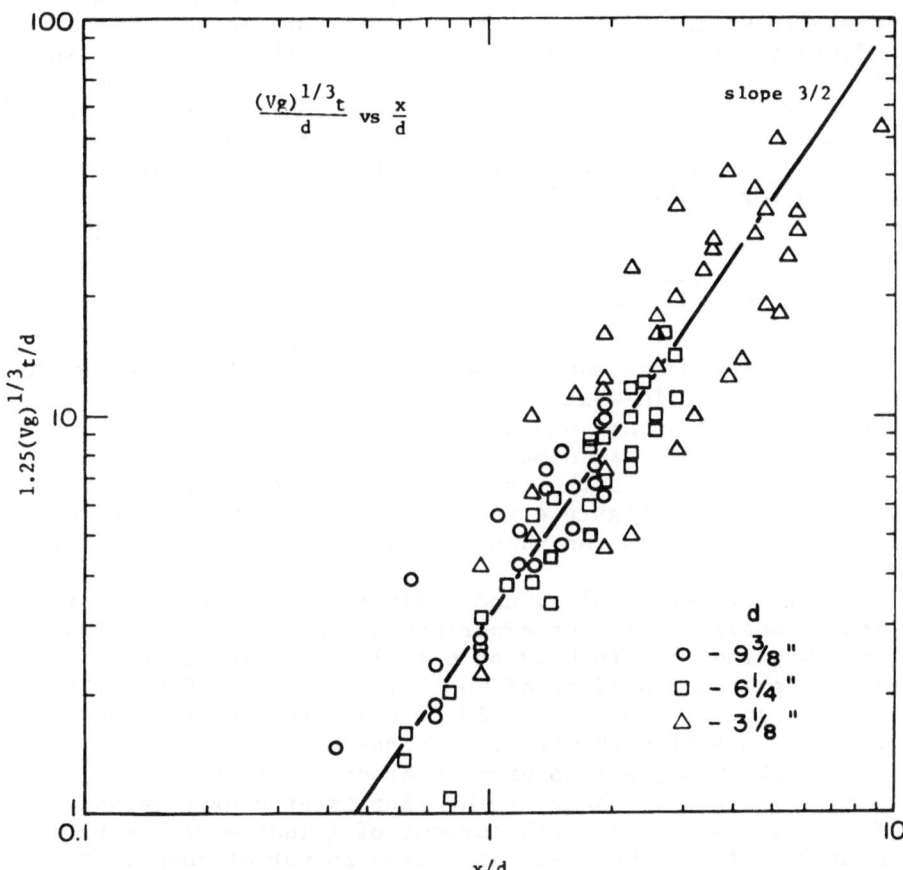

Figure 3. Comparison of equation(7) (the solid line) and data from Solomon (ref. 4).

it is in still water. However, if the oil is drifting
towards the boom with a velocity U_D, the surface current
generated by the boom must be strong enough to stop the oil
a distance from the boom, for otherwise, the orbital motion
of the wave field will carry a particle across the boom.
Increasing V, the volume rate of flow, increases the stand-
off distance of the oil, x_s. According to the above assumptions, x_s is the distance at which $U = U_D$, and equation (5),
derived for still water, may be used to find

$$\frac{x_s}{d} \sim \frac{(Vg)^{2/3}}{U_D^2} \ . \tag{8}$$

To check these assumptions, an experiment[4] was run in
our laboratory, in, which U_D was the drift velocity due to
the wave field alone, and no wind or tidal currents were
present. The oil was simulated by small wood chips. The
major uncertainty in the experiment was the accurate deter-
mination of U_D. Figure 4 shows the results, which are in
fair agreement with the theory.

Using these results, and evaluating the constants of
proportionality from the experimental data, we may estimate
the power required to hold an oil slick. A reasonable
power might be 1 hp/foot of boom, i.e. about 5000 hp/mile of
boom. Suppose, the boom is 12 ft below the surface, and
operates in 4 ft high waves. A tidal current pushs the oil
against the boom, but no wind is present. In this case, the
boom will leak oil whenever the tide is greater than about
1/3 knot. To stop a tidal current of 1 knot would require
about 30 hp/ft of boom--an excessive amount of power. This
simple example serves to show that <u>the pneumatic boom in its
present form is limited by the power available.</u>

III. PHYSICAL BOOMS

A physical boom is designed to produce a barrier,
usually vertical, to mechanically stop the flow of oil. Let
the depth of oil which may be contained at the barrier be d.
Because the barrier may not remain vertical as it moves in a
wave field, and is deflected by currents, we expect d to be
considerably less than the draft of a floating boom in still
water. At present, the relationship between the properties
of the boom, the wave field and currents, and the effective

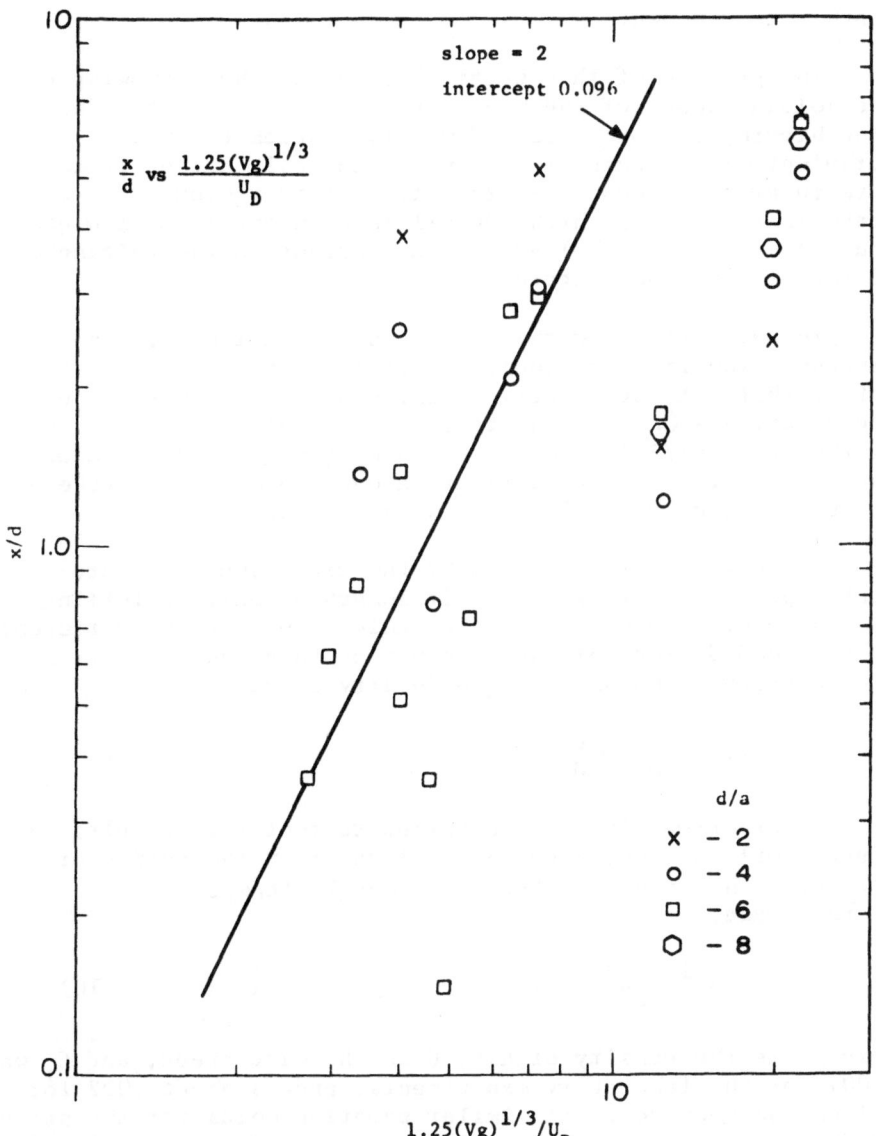

Figure 4. Comparison of equation(8)   (the solid line) with data from Solomon (ref. 4).

depth d is unknown.

The purpose of this section is to show how, knowing d, the holding power of the boom may be estimated. Consider a wind blowing over the oil and water, towards the boom. The turbulent wind stress, τ, acting on the oil, causes it to move towards the boom. If the oil is very viscous, a steady state is reached in which the oil is essentially motionless, and the stress τ is balanced by a gradient in the thickness of the oil, h. See Figure 5.

To understand how the thickness, h, varies with the distance away from the boom, x, refer to Figure 6. First notice that, if the relative density difference between oil and water, $\Delta = (\rho_w - \rho_0)/\rho_0$, is small (a typical value for oil on the sea is 1/10), then $(1 - \Delta)h$ of the oil floats below the water surface, and Δh floats above the surface. Like an iceburg, 90% of the oil is below the surface.

The stress τ is balanced by the difference in hydro-static pressure acting on a unit length of oil. Referring to Figure 6, it can be seen that this is simply the difference in the shaded areas of the pressure-depth graphs shown. A simple calculation shows (ρ_w = density of water)

$$-\tau = \rho_w g \Delta h \frac{dh}{dx} \quad . \tag{9}$$

An experiment[5] was performed to test this result. A viscous oil was contained in still water by the action of a barrier. The wind velocity was about 10 ft/sec. In this case,

$$\tau = \frac{1}{2} \rho_a U^2 C_f \quad . \tag{10}$$

Here ρ_a is the density of air, U is the wind speed, and C_f was 0.007 for the laboratory experiments, and is about .002 for oil on the open sea. (A similar equation holds for the stress produced by a current in the water, provided ρ_a is replaced by ρ_w.)

The experimental data consists of a measurement of the mean slope of the oil at the barrier, for various thicknesses, h, and wind speeds, U. Figure 7 shows this data compared with the prediction of equation 9. It is seen that there is

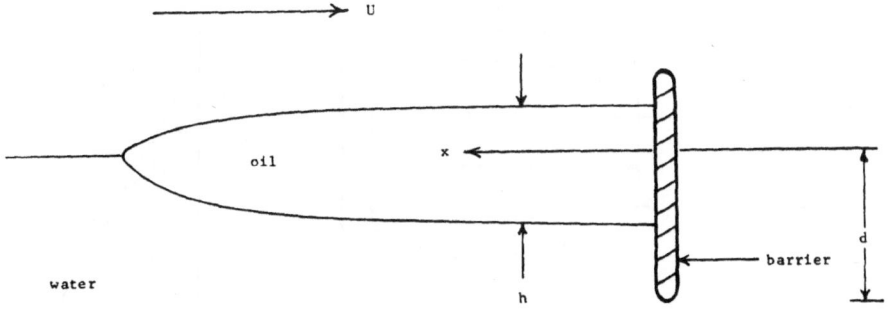

Figure 5. The wind, with velocity U, pushes oil against a
barrier of depth d. The wind-induced turbulent stress,
acting on the upper surface of the oil, pushes the oil
against the barrier.

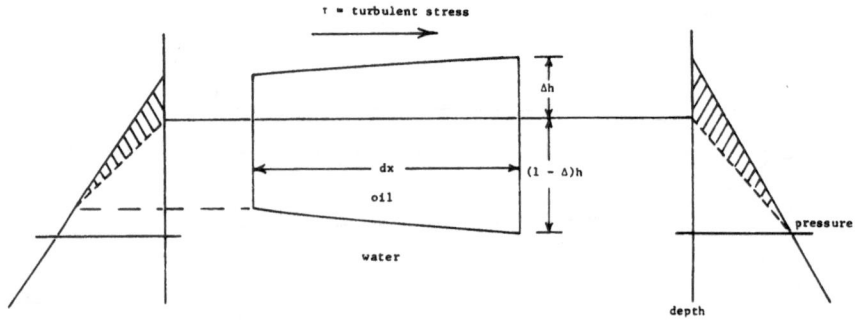

Figure 6 shows how the turbulent stress, acting on the upper surface of the oil, is balanced by the gradient in the thickness of the oil, h(x).

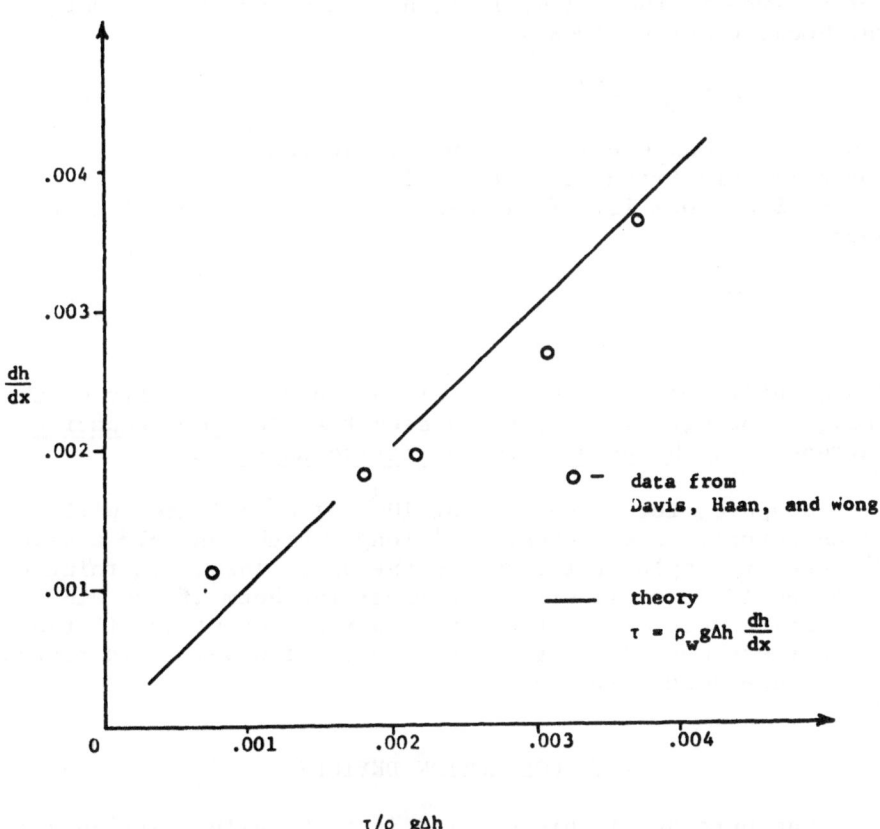

Figure 7. Comparison of the theory of equation (7) (the solid line) with the data of ref. (5).

good agreement between experiment and theory.

We may use this initial slope to estimate the holding power of the boom. Suppose the thickness of the oil at the boom is equal to the effective depth of the boom, d. Then for a given τ, the oil will extend a distance L in front of the boom, L being given by

$$L = \rho_w g \Delta d^2 / 2\tau \quad . \tag{11}$$

Since the cross sectional shape of the contained oil is approximately triangular, the volume Q oil contained by a length ℓ of boom is, if the wind direction is normal to the boom,

$$\frac{Q}{\ell} = (\frac{1}{2}) \frac{\rho_w g \Delta d^3}{\rho_a U^2 C_f} \quad . \tag{12}$$

The quantity Q/ℓ is a measure of the holding capacity of the boom. From equation 12, is is seen that the <u>boom capacity increases as the cube of the effective depth, d.</u>

Suppose, for example, that 10^5 gal of oil are spilled in an estuary, and a boom, 200' long, which can hold a slick 3" deep, is deployed to contain the oil. Then, according to equation 12, the oil will leak under the boom if the wind velocity exceeds 12 knots with no waves present, or if the current exceeds 1/3 knot. The effects of waves would reduce these threshold values.

IV.   COLLECTION DEVICES

The purpose of this section is to briefly describe some of the collection devices currently in use. One such system is the "Port Service" built by Surface Separator Systems, Inc., and in use in Baltimore, Maryland. This vessel is 38'-6" overall, has a beam of 16' - 6 1/4", and a draft of 3'-9". The vessel can carry 3000 US Gal of collected oil, and is powered by a 52 hp diesel.

In the port of this barge-like vessel are a series of absorbent rollers, half immersed in water. These rollers preferentially absorb oil on the surface of the water. As the roller is rotated slowly, the oil is lifted out of the

water, and then squeezed out of the roller. The quoted
collection rate is 830 gal of oil/hour. A disadvantage of
this system is, that it does not work in waves higher than
9", and so is appropriate for use only in well sheltered
harbors. The "Port Service" cost is quoted as 10^5.

A less elaborate system, operating on the same principles,
is built by American Oil Corp. It consists of a 24' long
catamaran, powered by an outboard motor, with the roller
between the hulks. The roller consists of a 4' long,
1' diameter drum coated with hydrophobic polyurethane foam.
The system can collect up to 1500 gal/hour. Again, this
system can operate only in waves less than 6" high.

A few facts may be raised to put these collection rates
in perspective. To unload a supertanker in the usual time
of 3 hours, the oil flow rate is about 500 gal/sec. For
every second of oil spillage at this rate, 30 minutes of
collection time would be required, using present devices.
Again, using 20 such devices, 100 days would be required to
collect the oil from the Torrey Canyon.

In summary, it is clear that present collection systems,
which are designed to collect small amounts of oil spilled
in still water, are inappropriate for large spills in exposed
areas.

REFERENCES

1. G. M. Hidy and E. J. Plate, Journal of Fluid Mechanics,
    26, (1966), 651.

2. See Dynamics of The Upper Ocean,
    O. M. Phillips, Cambridge University Press, 1966,
    pp 31-43.

3. This result was first obtained by G. I. Taylor, Proc.
    Roy. Soc. A, 231, (1955), 466.

4. L. Solomon, "Pneumatic Boom for Containing Oil Slicks
    on The Sea", M.S. Thesis, Department of Mechanical
    Engineering, M.I.T., 1969.

5. S. Davis, A. Haan, and J. Wong, private communication.

REMOVAL OF FLOATING OIL SLICKS BY THE CONTROLLED
COMBUSTION TECHNIQUE

PAUL R. TULLY

SECTION HEAD, (CAB-O-SIL RESEARCH AND
DEVELOPMENT) BILLERICA LABORATORIES
CABOT CORPORATION, BOSTON, MASSACHUSETTS

In March of 1967 an event occurred which has spurred
intensive worldwide research efforts directed towards
solution of the problem of removal of oil spilled at sea
and on inland waters. The Torrey Canyon, a heavily laden
super-tanker, grounded off the southwest coast of England,
broke up, and disgorged about 850,000 barrels of crude
oil onto the open seas. Despite intensive corrective
efforts by governmental and private groups, large
quantities of oil washed ashore on Cornwall in England
and Brittany in France.

Subsequently, the Ocean Eagle hit bottom off San Juan,
Puerto Rico, and four days later the General Colocotronis
beached near Eleuthera Island in the Bahamas. Both
released their cargoes of crude oil onto nearby beaches.
In the winter of 1968-69, large oil slicks caused varying
degrees of damage to the shorelines at Santa Barbara,
California, and to the coasts of Delaware and Connecticut.
Many smaller slicks are continually reported. The resulting pollution of the seas and the land has had and will
continue to have widespread and long-term detrimental
effects on the wildlife and economics of the afflicted
areas.

Accordingly, there has been a focusing of interest
on providing suitable remedies. To date, three general
methods of dealing with the problem have been explored
by industry and government agencies. These are:

81

absorption, followed by removal or sinking; the use of
chemical dispersants; and removal by burning.

Absorption and sinking techniques rely essentially
on the absorption of oil by clays, expanded micas,
diatomaceous earth, talcs, and similar materials. The oil
is absorbed by the solid material and the resulting mass
is then physically removed or allowed to settle to the
bottom of the body of water.

Such treatment methods suffer several disadvantages.
First, the amount of solid material required is generally
very high, e.g., 20 to 50 per cent or more by weight of the
oil. Second, application does not usually confer
sufficient additional physical integrity to the oil mass
to render a subsequent physical removal step practical.
When, on the other hand, such absorbed slicks are allowed
to sink to the bottom of the water body, their eventual
elimination is dependent almost entirely upon biodegradation
process and/or the dispersing action of waves, tides and
ocean currents. These latter processes, in turn, may cause
the oil to reappear on the surface at a later date and
eventually wash ashore. In addition, the presence of large
submerged oil masses represents a continuing threat to
ocean life and to the fishing industry.

The second method of dealing with oil slicks depends
upon the treatment of the oil slick with a dispersing
agent, such as an emulsifier, detergent, or surfactant.
The dispersant acts to break up the oil slick under shear
action into relatively small globules which then sink and
are conveyed to the bulk regions of the water. The
dispersed globules of oil then biodegrade with time.

Effective dispersant materials, however, are often
detrimental to many forms of wildlife indigenous to a
marine environment. Further, this method has been relatively ineffective for the removal of thick slicks of crude oil,
the type that is most common and most devastating. The
shear forces necessary to bring about physical breakdown
of the slick and combination with the dispersant chemical
are large. In the absence of vigorous wave motion these
forces are not readily achieved by such mechanical means
as a boat's wake action.

The ultimate solution of the oil slick problem, then,

would seem to lie in the removal of the oil from the afflicted water mass rather than mere oil dispersal. The third major method proposed does involve removal of the oil--by burning it from the surface of the sea. This method eliminates essentially all the spillage. Historically, this apparently simple process has proved to be ineffective. Heavy oil, in particular, has been found impossible to ignite. Even when a slick is forcefully ignited (for example with the aid of flame throwers) combustion cannot be maintained; the fire quickly dies out. In the case of the <u>Torrey Canyon</u> disaster, the addition of thousands of gallons of aviation fuel and Napalm combined with aerial bombing of the ship failed to produce any sustained burning.

The difficulty experienced in burning oil slicks is associated with the rapid loss by evaporation of the more volatile components of the oil with time while on the water. Also when combusted there is a transfer of heat to the underlying water mass, decreasing the oil temperature to below the flash point[1],[2].

Research personnel of the CABOT CORPORATION, a company long engaged in the manufacture of fine particle materials for industry, became intrigued by the possibility of using these materials to create a wicking mechanism which would permit the oil to be burned. As a result of a research program, we developed a new product called CAB-O-SIL ® ST-2-0, and a practical method of application of the material to the surface of the slick using standard readily available fire-fighting equipment. Using CAB-O-SIL ST-2-0 and the recommended treatment methods, approximately 98 per cent of such heavy oils as crude and Bunker C can be effectively burned off the water. After burning, a non-tacky, hardened residue, similar in appearance to tar paper, remains in a readily collectible form.

---

(1)   Battelle Memorial Institute, Oil Spillage Study, AD 666289 IV, Burning.

(2)   Burgoyne, J. H., A. F. Roberts and P. G. Quinton, 1968 Proceedings of the Royal Society, Series A 308, p. 52.

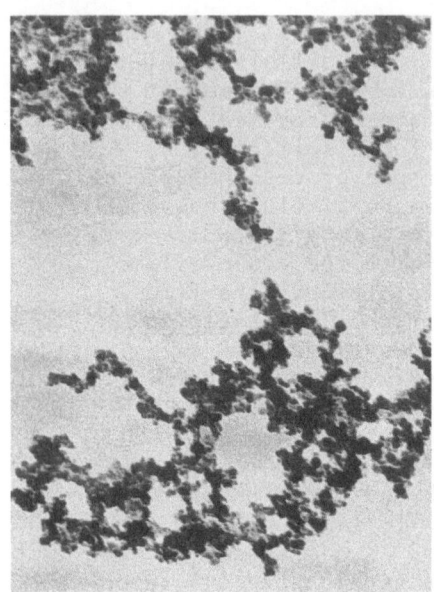 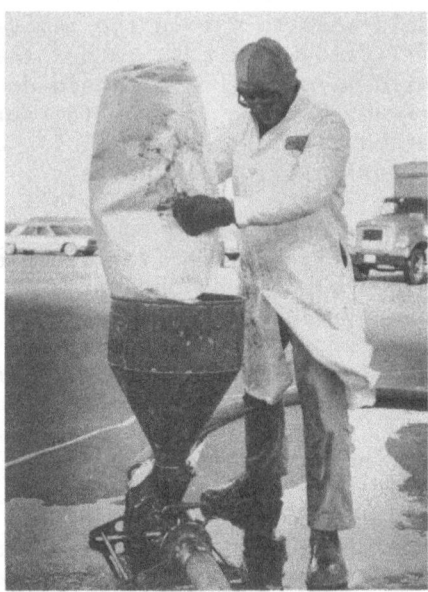

TOP LEFT: Unusual structure of Cab-0-Sil particle is visible in electron photograph of 50,000 magnification.

TOP RIGHT: In test, Cab-0-Sil ST-2-0 is introduced into water stream through standard foam generator.

BOTTOM: Mixture of Cab-0-Sil ST-2-0 and water is sprayed onto surface of floating oil.

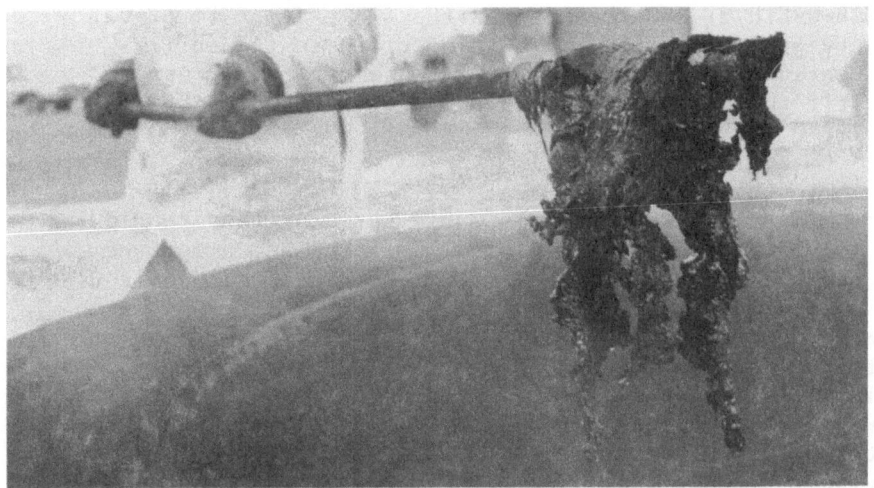

TOP: As test continues, foam-like Cab-0-Sil on oil sur-face promotes spreading of flame front. In less than ten minutes later, flame covered entire treated surface.

BOTTOM: After combustion, crusty residue of about two percent of original oil volume remains and is easily scooped from water.

CAB-O-SIL ST-2-O is composed of extremely fine
particles of fumed silica surface-treated with a silane
coating to render it hydrophobic. The material is non-
toxic. The method of application consists of covering
selected areas of the oil slick surface with a layer of
the material by entraining the product in a stream of
water which conveys it to the surface of the slick.

As soon as the mixture strikes the oil, the water
and the hydrophobic CAB-O-SIL ST-2-O separate. The water
sinks under the oil, and the CAB-O-SIL ST-2-O rises to the
top of it, forming a thin, foam-like coating in that
region of the slick where combustion is desired.

The treated area can then be ignited. Ignition is
conveniently accomplished with a piece of cloth which has
been saturated with lighter fluid, then dropped onto the
treated surface and ignited. The wicking action of the
CAB-O-SIL ST-2-O then causes the flame front to travel
very gradually from the burning cloth to the entire
treated area of the slick and sustains the combustion
until the oil in the immediate area of the CAB-O-SIL
ST-2-O is consumed.

In the course of our research effort, a great
many particulate materials whose surface areas ranged
from 0.5 to 400 square meters were studied, among them
asbestos, talcs, silicas, micas and clays. The oils used
in these experiments were No. 5 and 6, neither of which
will ignite and burn while floating upon water without
the use of promoters. Though some particulate materials
initially floated and supported initial combustion, they
were of such density that their large heavy nature caused
them to sink into the oil as burning reduced the oil
viscosity, and the fire was extinguished.

CAB-O-SIL ST-2-O, however, did not sink through the
heated oil layer but effectively promoted combustion of
the oil, as indicated in the diagram. The CAB-O-SIL
ST-2-O floats because of its low bulk density, (2 - 3
pounds per cubic foot). A fluffy powder, CAB-O-SIL ST-2-O
is so open in structure that only 10 pounds of material
occupies a multi-wall bag normally used for 100 pounds of
common industrial minerals. In addition, it is theorized
that the extraordinary thickening action of CAB-O-SIL
ST-2-O, its chief characteristic when used in industrial

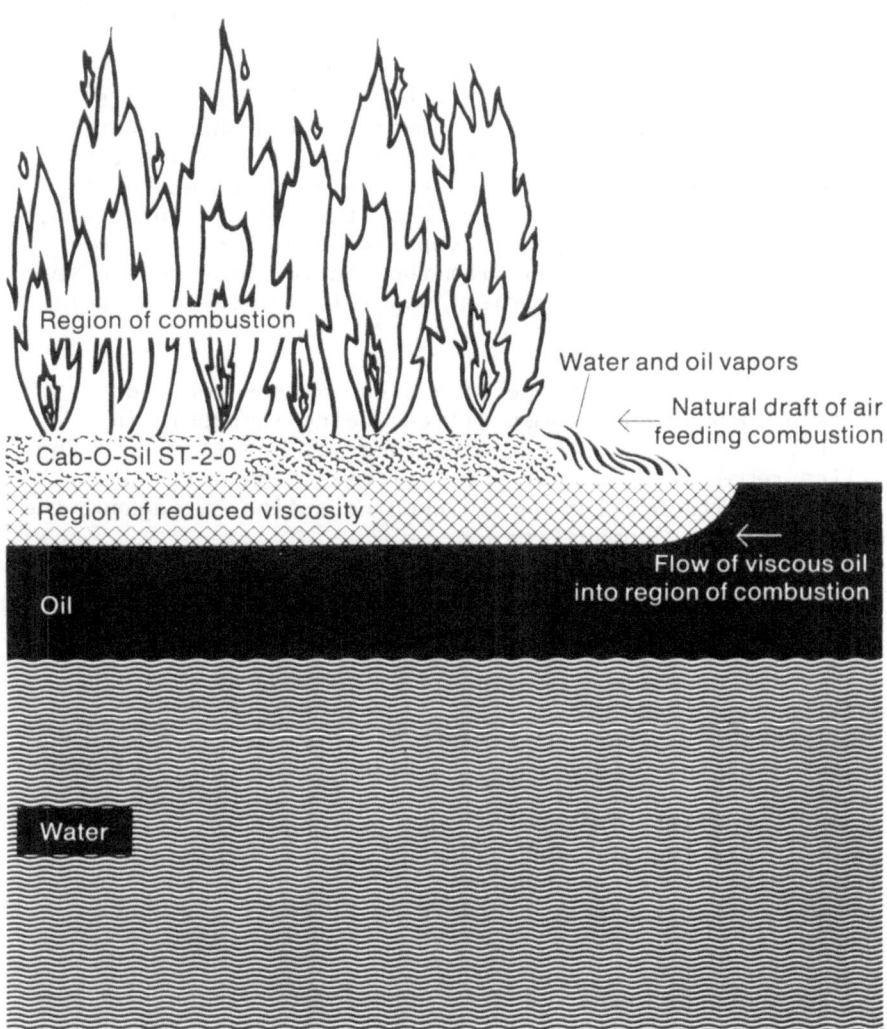

Diagram illustrates the mechanism of controlled combustion assisted by Cab-0-Sil ST-2-0 on floating slicks of heavy oil. Acting as a wick, Cab-0-Sil on surface draws up oil by capillary action, increasing the rate of vaporization. Burning of surface oil heats oil below it, reducing viscosity and facilitating further wicking. At edge of burning, oil and air flow towards flame.

processes, may be acting to thicken the heated oil just
below the surface, contributing to buoyancy.

The wicking action of the CAB-O-SIL ST-2-0 appears to
be a specific example of the common wicking phenomenon.
In this case, a layer of fine CAB-O-SIL ST-2-0 with tiny
channels among the particles, acts as an excellent wick.
As a wick, it increases the area of the oil-air interface,
thus increasing the rate of evaporation of the oil. As
the mass transport of oil from the liquid phase to the vapor
phase is increased, sufficient vapor is formed to support
combustion.

The continued presence of a wick is essential to
sustain combustion. Fumed silica, being inorganic in
nature, is immune to the temperatures of an oil fire. The
silica remains at or near the surface of the oil,
continuing its wicking and thickening action throughout the
duration of the combustion.

The need for a wick to support combustion has another
important aspect. All our experience indicates that the
oil will burn where the CAB-O-SIL ST-2-0 is applied, and
will not burn where no fumed silica exists, despite the
heating of adjacent oil by the fierce heat at the edge of
the fire zone. The significance of this fact is that in an
actual emergency operation, the size of the fire zone can
be accurately controlled by limiting the initial spreading
of the material. The operational personnel, if desired,
can burn off the oil zone by zone, always keeping the fire
under surveillance and control. Additionally, the fire is
quickly extinguished by standard water fog spray where
needed.

As the fire burns, all the dynamics of oil burning are
operating. The oil directly beneath the flame, and in an
area extending slightly beyond the edge of the fumed silica
blanket, is warmed by the heat of the fire, reducing the
viscosity of the oil. This reduction of viscosity
accelerates burning once the fire gets underway, because
oil of low viscosity wicks and volatilizes more rapidly.

The elimination of oil by combustion produces a
depression in the surface of the oil, creating a "hole"
into which surrounding oil flows. The in-flowing of the
oil towards the fire precludes the need of applying

CAB-O-SIL ST-2-O to the entire slick, permitting economical utilization of the material and thus simplifying the logistics of conveying the material to the scene.

Above the burning oil surface, the upward movement of the heated air creates a vacuum into which cold surface air rushes. The inward-flowing air fans adjacent vapors into the flame. This tends to centralize the fire and prevent spreading of the flames to the volatile oil vapors formed beyond the area of the immediate fire.

Another action that is believed to be in process in this area of reduced viscosity, as indicated in the diagram, is the evaporation of imbibed water. Floating oil imbibes water over a period of time[3], further reducing combustibility. But the heated oil in the region of reduced viscosity evaporates contained water, increasing the combustibility of the oil about to be burned.

Laboratory work shows that the method of burning with the assistance of fumed silica is effective with slicks of any thickness down to about two millimeters. The quantity of CAB-O-SIL ST-2-O required varies with the slick thickness, nature of the oil, and conditions of application. Recommended quantities are from 0.1 to .5 per cent, based on weight of CAB-O-SIL ST-2-O to weight of oil.

CABOT's purpose has been to develop a material that would work not only in theory, but could be applied practically and economically under conditions at sea. Since any dry powder applied directly to the oil would drift uncontrollably with the wind, we delivered a factory process for rendering the CAB-O-SIL particles

---

[3]   S. A. Berridge, M. T. Thew and A. G. Loriston - Clarke. "The Formation and Stability of Emulsions of Water in Crude Petroleum and Similar Stocks", Jour. of the Inst. of Pet., Nov., 1968.

TOP: In test tank at the Moon Island Research and Train-
ing Academy of the Boston Fire Department, entire Cab-0-
Sil treated surface of oil burns fiercely until 98 percent of
the oil is destroyed.

BOTTOM: Emergency application of Cab-0-Sil ST-2-0 to
2000-gallon oil spill at Heard Pond, Wayland, Massachusetts, removed all the oil from the pond, preventing
pollution of a federal wildlife refuge.

hydrophobic by surface treatment (hence the "ST" designation) with silane, a water repellent material. The surface treatment keeps the fumed silica particle essentially dry while they are being entrained into a water stream.

Although the basics of application and ignition are simple, an oil slick eradication operation with fumed silica must be conducted by personnel who understand the action of the material and the action of combustion processes, and who are trained in standard fire fighting methods. These requirements would be met by professional fire-fighting groups working in conjunction with CABOT personnel or having previously been counseled by CABOT personnel.

We are well aware, of course, that the oil combustion produces a smoke column which is a form of pollution itself. Given the present inadequacy of most proposed solutions to the oil slick problem, however, and the high probability that despite precautions oil spills will continue to occur for years to come, it is felt the total damage to the environment by the quickly dissipated smoke is only a small fraction of that which is done by oil in the water. This view is shared by conservation groups.

This new material became commercially available in January of this year. It is the subject of interest on the part of government agencies, oil company personnel, insurance companies and conservationists. Active evaluation is underway in several locations.

As legislation enacted by the federal government and state governments clarifies the lines of authority for quick action in oil slick emergencies, we believe that the responsible governmental and private agencies will arm themselves with the knowledge and equipment needed to use the assisted-burning method of oil slick removal wherever necessary.

OIL TRANSPORTATION BY SEA

R. F. COOKE

GULF OIL TRADING COMPANY
1290 AVENUE OF THE AMERICAS
NEW YORK, NEW YORK 10019

I am pleased to have been invited to discuss today transportation of petroleum by sea. I cannot, of course, deal comprehensively with this subject, but will touch upon some facts, and problems in the minds of tanker and oil industry people today. This is the 25 cent survey course.

The acceleration of tempo in the oil tanker business during recent years has been extraordinary, especially when viewed in the light of earlier, strong conservatism. This activity has been stimulated by the oil companies' need to remain competitive in a very competitive business, and indeed if possible, to set the pace. The drive in recent years to reduce costs in the oil industry has pointed up the important role played by transportation in determining profitability of an enterprise.

The steadily expanding economies of the industrial nations, and the emergence of so many underdeveloped countries has resulted in a growing demand for the materials of industry and for a higher standard of living. Petroleum being one of the industrial staples, this industry has experienced the large increase in demand, and by worldwide exploration and subsequent production, is continuing to meet this demand. Many new refineries have been and are being built or enlarged. These efforts result in sufficient availability of oil, which, however,

also must be moved from production source to market.
Although to some extent, the discoveries of crude oil in
new locations have tended to shorten the average distance
the oil must be carried, the increased volumes from
Middle East sources outweigh this tendency and result in
net increased requirements for world tanker transportation.

Considering the present trands in the tanker industry,
it is probable that developments at the Suez Canal will
have very little effect on transportation of oil. With
the advent of extremely large size tanker tonnage, oil
companies are becoming more and more independent of the
canal for economy and are tending to provide sufficient
tonnage to supply their needs via the Cape of Good Hope.

In order to focus upon the magnitude of oil volumes
being moved by sea, let us look at the situation over the
past 20 years. A parameter for measurement of this transportation is the ton mile per unit time, or in other words
the vessel carrying capacity. This is commonly expressed
in terms of "T2 equivalents," or the number of standard
T2-SE-A1 tankers, wartime built, which would be required
to perform a given number of ton miles per unit time. The
total number of T2 equivalents in the world tanker fleet
is projected at 11,000 in 1970. Over the period of 20 years,
capacity has increased more than sixfold, and has approximately doubled since 1960. The curve appears to have an
exponential trend. With a total increase of 2,200 vessels
estimated for the period 1969 through 1972, you can see
this is a subject for much interest and discussion and
speculation.

Another way of looking at these volumes is the amount
of tonnage trading. The curve represents total deadweight
tonnage operating over this same period. On the average,
about 40% of total tonnage is loaded with cargo at any
given time. Therefore, at this moment, there should be
about 50 million tons of oil plying the waters of the world.

It is sobering and very pertinent to consider that had
the industry not increased tanker sizes beyond that of the
wartime T2, the congestion in our shipping lanes would by
now have become intolerable and probably collisions would
have become a common occurrence in the sea lanes of highest
concentration close to densely populated areas.

A total of 3,500 tankers were trading during the 1966 period. Had all vessels been of the T-2 size there would have been a total number of 6,600 vessels traded in that period. It is estimated in 1970 3,800 tankers will be traded. T-2 equivalents in 1970 will total 10,400 vessels.

It is highly problematical in the minds of tanker people, whether there is more or less risk involved in the use of larger tankers.

Ask yourself, for example:

Is a large ship more likely to be stranded than a small one? Are two 200,000-ton vessels more likely to collide than forty 15,000-ton ships?--- This, of course, is equivalent tonnage.

While pondering these questions, bear in mind that regardless of tanker size, the same total quantity of oil is to be carried, and thus the same total quantity is on board tankers at sea at any given time.

The above considerations alone would lead to over-simplification. However, it may serve to illustrate that the coin has two sides.

Probably the most dramatic and certainly by far the most widely publicized trend in tanker construction has been that of tremendous increase in sheer size. It is interesting to consider the growth in size of tankers over this same period of time.

Historically, maximum tanker size has progressed from under 20,000 dwt. of 1930 to a projected 800,000 dwt. sometime in the early 1970s. Again, there has been an exponential trend.

The largest vessels to date are the six 327,000-ton tankers constructed in Japan; four of which are now successfully operating. However, there is no real expectation that these ships will hold the title very long. Reliable sources indicate that one of our competitors is presently negotiating for a class of tankers approximating 400,000 dwt. Studies have been undertaken by Classification Societies on 500,000 dwt. tankers and undoubtedly vessels of these sizes will soon be contracted.

In order to realize the relative importance of ocean
transportation on the cost of petroleum, let us assume
the price obtainable per barrel of oil is $2.00 at the
destination of a 3,000-mile (one way) voyage. Depending
upon size of vessel transporting this cargo from source
to destination, the profit margin (or loss) may be sub-
jected to a variation of about 46% due to freight alone.
It may be implied from the aforementioned that, generally,
larger ships mean lower freight cost, all other conditions
being similar.

Mention of large tankers in any group can hardly be
made without raising the spectre of the TORREY CANYON
disaster. The subject is a serious one, and is considered
so by all responsible people in the industry. However,
the problem is one which has been with us for a long time.
Solutions to the largest part of the water pollution problems have been worked out by the oil companies in recent
years, in the "Load on Top" system, which is widely used,
but not universally accepted. It is to be hoped that progress in the industry will not be hampered by unduly re-
strictive regulations as an emotional aftermath of this
incident.

What is required is ever-increasing vigilance over
tanker movements and anti-pollution measures on the part
of oil companies and independent tanker owners, otherwise
we may be sure that the initiative will be taken out of
our hands by government. This would be a major set-back
and would admit a grave deficiency in the ability of
private enterprise to carry out its business in an orderly,
safe and responsible fashion.

Complete success in preventing oil pollution can only
be achieved when all tanker operators can be persuaded to
adopt the "Clean Seas Code," and a most interesting development in this connection is a voluntary scheme proposed and
announced by seven Major Oil Companies whereby national
Governments would be compensated for the cost of cleaning
up oil spillage off their coastlines. The scheme will be
known as the "Tanker Owners Voluntary Agreement concerning
Liability for Oil Pollution" (TOVALOP). Under this scheme
the Owners would be responsible for cleaning up as a result

of a marine incident, attributable to the fault of the vessel but if they are unable to arrange for the cleaning, provision is made for reimbursing the National Government to cover its reasonable costs of the cleaning-up operation.

All tanker owners, Private, Oil Company and Government, will be invited to participate in the plan which would become effective when joined by companies representing 50% of the World's tanker tonnage. The seven sponsors represent about 30%. The plan, once operative, would remain in force for a minimum of 5 years to be followed by successive two-year periods. It will, however, lapse after two years if by that time tanker owners representing 80% of the World's tanker tonnage have not joined.

I should like to give a brief description of the Bantry Bay operation which commenced about six months ago. The transshipping terminal is located on Whiddy Island, near the head of Bantry Bay, a large bay in the southern coast of Ireland, which at one time was an anchorage for the British Royal Navy.

The main requirement for the terminal was very deep water and the ability to be able to operate safely and efficiently throughout the year. The terminal had, too, to be as closely equidistant as possible to our refineries. Another very important consideration, was that of safety. We were planning the Bantry Project well before the "TORREY CANYON" disaster; yet it was quite clear to us at the time that, with tankers of the size envisaged, it was imperative that their operating route should be as far as possible from the principal shipping lanes generally in use to and from Europe, and well clear of all navigational hazards.

With a maximum water depth of 80 ft. which progresses almost up to the shore it was possible to locate an island type dock very near to the shore while still permitting a 100,000 dwt. vessel to dock on the inshore side. A good approach channel is also available, with marker buoys installed as navigational aids.

The shore installation consists basically of twelve 600,000 barrel crude tanks of floating roof design and two 500,000-bbl. dirty ballast tanks of similar structure. Tankage is also provided for bunkering purposes. Other ancillary services, of course, are included.

Two 42-in. submarine crude oil lines and one 30-in.
dirty ballast line are manifolded on the dock and inter-
connected by crossovers to inboard and outboard articu-
lated pipe arms. The design capacity of the system is
75,000 barrels per hour for crude oil and 50,000 barrels
per hour for ballast at each berth.

In order to relieve the stress on the dock when a
300,000 dwt. vessel is in berth, and subject to high winds,
the minimum draft of these vessels at any time in berth
will be 40 ft. This, then, means that for a part of the
discharging operation of these vessels, ballast must be
taken on concurrently with the oil discharging in order to
avoid undesirable delay. The cargo and ballast piping
systems of the ships have been designed to perform in this
manner.

Shuttle ships arrive at Bantry Bay terminal from such
ports as Rotterdam (Europoort) and Milford Haven. Since
these voyages are very short, the shuttle ships do not have
sufficient time to clean tanks and arrive at Bantry with
only clean ballast. Therefore, they arrive with dirty
ballast, which they pump either ashore into the dirty
ballast tankage or directly across the dock into a 300,000
dwt. ship if the timing coincides. On a statistical average
basis, if the shuttle ships carry 40% of their deadweight
in a ballast condition, which is rather conservative, than
the volume of ballast to be taken by the large vessels
corresponds to about 44-ft. draft. If during any operation
insufficient dirty ballast is available from incoming
shuttle ships to maintain at least 40-ft. draft, the sup-
plementary ballast is taken from sea suction. If more
ballast is available, it is loaded, allowed to settle for
the period necessary, and part decanted as clean ballast at
sea. The residue of dirty ballast is passed through an
installed oil/water separator, after which the clean water
is pumped overboard and the oil residue collected in one of
the cargo tanks. Meanwhile, as cargo tanks become emptied
of dirty ballast, they are washed, using the installed tank
cleaning system, and filled as appropriate, with clean
ballast, constantly maintaining a normal ballast draft of
30 ft. In this manner, the vessel arrives at Kuwait with
clean ballast only, but takes on additional clean ballast

immediately prior to arrival in order to berth with a
minimum 40-ft. draft, as at Bantry, and for the same reason.

At Bantry, the most modern oil-spillage precautions
have been taken with the provision of a highly sophisti-
cated oil/water separation system, and oil boom and skimmers, as well as dispersal equipment on the four tugs and
at the terminal itself.

As far as the tankers are concerned, our answer has
been to ensure that the vessels are staffed by officers of
the very highest calibre and experience, and that they are
able to make use of the most advanced navigational and
other equipment available.

Considerable attention has been given to the handling
of these enormous ships by tugs and mooring equipment.
Three large tugs are available at Bantry at all times, each
capable of a sustained bollard pull of 30 tons. Each tug
is fitted with extensive fire fighting equipment, as well
as the oil spill dispersal equipment. Two of the tugs are
also fitted with derrick for handling marker buoys and
their ground tackle, which will be maintained by the terminal facilities.

Studies and discussion have produced a mooring system
on the ships which matches the reinforced strong points on
the piers at Bantry and Kuwait. The winches are of the
automatic constant tension type, having individual rendering capacities of up to 50 tons line pull.

The UNIVERSE IRELAND, KUWAIT, JAPAN and PORTUGAL,
four of the six 300,000 dwt. class ships contracted for and
now in service are indeed the largest ships afloat -- 1,135
ft. long by 175 ft. breadth by 105 ft. depth, with a deadweight of 327,000 tons and about 81'6" draft.

The machinery is twin screw, steam turbine, totalling
34,000 normal Shafthorsepower supplied by two boilers to
propel the ship at about 15 3/4 knots average service speed.

The machinery spaces, boiler and engine room are con-
servatively designed with a degree of automation, and controlled from an air conditioned central control room from

which the machinery can be observed.

There are 4 main cargo pumps, each having a capacity
of 3500 cubic meters per hour of sea water at 125 meters
total head. The cargo system is fully operable remotely
from a central cargo control room. Tank levels, valve
positions, and pump pressures can also be monitored from
this location.

The ship is fitted with an installed tank washing system, a gas-freeing system, and an oil/water separating system for the cargo tank spaces.

An effort will be made to prolong the usual 12-month
drydocking interval on these vessels. To this end, the
propeller shafts are fitted with white metal stern bearings
rather than the usual lignum vitae. Coatings have been
extensively applied in lieu of conventional paints, and
materials and equipment have been chosen for maintenance-
free performance.

Nevertheless, a time must come when these ships will
need to be drydocked. At the time of making the decision
on vessel size, there were only four known shipyards which
could dock vessels of this size. It is now anticipated that
10 or 12 such docks will be available (newbuilding and
repair).

The entire external hull is coated with shop primer,
inorganic zinc silicate, and cure coat. In addition, three
coats of vinyl are applied to the hull from keel to rail,
and antifouling from keel to deep loadline. The weather
decks are not coated over the zinc silicate.

The cargo tank space is divided into eight tanks fore
and aft, with two longitudinal bulkheads, making a total of
24 tank spaces. Although two of the wing tanks are desig-
nated as permanent ballast tanks, and handle clean ballast
only through a separate pumping system, a number of the
other tanks are coated internally (as well as the permanent
ballast tanks) and intended for carrying the dirty ballast
discussed previously. In addition to the preservation of
the steel, and ease of cleaning the coated surfaces, the
dirty ballast system is limited to these tanks, and a certain measure of control is thereby exercised over the hull

stresses induced by loading and ballast patterns.

The cargo tanks which are coated internally received a shop coat, inorganic zinc silicate and cure, with two overcoats of vinyl. Only upper and lower sections of the tanks, horizontal structural webs and cargo piping are coated, since these are the major areas of deterioration and tank cleaning difficulty.

Considerable care has been lavished on anti-pollution features, bearing in mind that this is extremely important at Bantry, a resort area. Therefore, in addition to the elaborate oil handling precuations already touched upon, facilities are also provided on board for the retention of sewage and garbage.

The main radio-telegraph equipment is single side band type, with a 120-watt emergency transmitter.

Ten-centimeter and three-centimeter radar units are provided, as well as Loran, Decca Navigators, and many other modern navigational aids.

It was interesting to me, when checking a few facts in preparation for this talk to discover that the tactical diameter, or turning circle for the 327,000 ton vessels experienced on trials, was about 2100 feet. A search in the archives revealed that the tactical diameter for a T2 tanker, 16,500 dwt., on trials was 2250 feet!

A further comparison between these two classes of ships, however, shows that the T2 performing a crash stop came to a standstill in the water in four minutes and 41 seconds, having travelled 3550 feet from the time of order, whereas the 327,000 tonner stopped in 10 minutes and 52 seconds, travelling 9450 feet.

Therefore, although the turning ability of these hugh ships is extraordinary, the master must "plan ahead," allowing for the mile and a half required to stop the ship from full speed condition. Of course, when maneuvering in restricted waters, and crowded sea lanes, the ship is not proceeding at full speed.

In order to cover so much territory it has been necessary to jump around quite a bit and to hurry through the
various points without dwelling on too much discussion or
explanation, but I sincerely hope the foregoing will give
all of you a general picture of oil transportation by sea.

THE ROLE OF THE FEDERAL GOVERNMENT IN CONTROLLING OIL POLLUTION AT SEA

Max N. Edwards

Partner, Collier, Shannon, Rill and Edwards
Washington, D.C.
Formerly, Assistant Secretary for Water Quality
and Research, U.S. Department of the Interior

The role of the Federal Government in preventing and controlling oil pollution of the sea has taken the form of Congressional and Administrative action, judicial interpretation, and participation in International Treaties and Conventions. These comments are an effort to examine some of the problems and touch briefly upon existing authorities to control this form of pollution. A look will also be given to pending Congressional action.

Oil pollution of the seas is both a national problem for all coastal states as well as an international problem. Pollution of this nature respects no national boundaries. Oil dumped into the high seas can slosh into the internal waters of any coastal nation and onto its beaches. Vessels constructed in one country, registered in another, and owned in still another move in and out of ports around the globe. Some of these tankers are safe and carefully run, yet seldom controlled by law. Others are obsolete and negligently operated by crews of questionable skill and experience.

This nation, and indeed the coastal nations of the world, must anticipate and be prepared to cope with spills even larger than the Torrey Canyon disaster of 1967, which broke up after grounding some eight miles from one of the Scilly Isles off Great Britain. It is interesting to note here that the Torrey Canyon was a "jumboized"[1] tanker carrying

---

[1]A term adopted by the industry which means enlarging the cargo-carrying capacity of an existing tanker.

over 119,000 (Dead Weight Tons) of crude oil from the
Persian Gulf to Milford Haven in Wales. She was 975 feet in
length, 69 feet in depth, and 125 feet wide. Once stranded
it was decided, for better or worse, to bomb the tanker from
aircraft rather than make any further effort to salvage the
vessel. When this happened thirty-five million gallons of
crude spread along the coasts of Cornwall and eventually
sloshed over 200 miles south and eastward to the beaches of
Normandy and Brittany.

Oil pollution of the sea is an international hazard not
likely to disappear in the foreseeable future. The demand
for ocean transported petroleum is expected to increase
throughout the world and the technology to achieve that
demand will be enlarged accordingly. But there can be no
guarantee against the massive oil spill any more than absolute
insurance of the faultless pilot or the perfect airplane.

There are many reasons why we Americans should concern
ourselves with preventing and controlling oil pollution.
One million tons of petroleum products move up and down our
coasts and inland waters each day, past valuable estuaries,
fish spawning grounds, resorts and vacation beaches. An
estimated 2,000 oil spills occurred in our waters in 1966
alone. We have enacted several laws to control oil pollution
in our Nation, in addition to those provisions of the International Convention for the Prevention of Pollution of the
Seas by Oil, 1954. It is clear, though, in the United States,
just as it is clear in the international arena, that these
laws are not sufficiently effective.

Let us consider the Torrey Canyon again as the classic
example of the international complexities of oil pollution.
Five nations are involved. The Barracuda Tanker Company of
Bermuda, a subsidiary of Union Oil Company of Los Angeles,
California, owned the vessel. For tax purposes, the ship was
leased by the subsidiary to the parent company, and it was
registered in Monrovia and flew the Liberian flag. The crew
was Greek and she was on a charter to the British Petroleum
Company.

The legal complications are a lawyer's field day. The
vessel was not British and crashed outside British territorial
waters, beyond the area in which that government had authority
to act. But escaping oil threatened the British realm. To
complicate matters, the vessel was not abandoned, in the

sense of maritime law, for some time, so the British government could not legally take charge of the situation.

Before it was determined who had legal authority to deal
in an effective way with the Torrey Canyon, two-thirds of the
oil had escaped.

Responsibility for the damages caused by the tanker
presented a unique number of legal questions concerning
venue, jurisdiction, recourse and the extent of damages.

When the British government sought to recover from the
tanker owners it was not clear where liability rested or
where suit should be filed. An action could be brought in
the United States (where Union Oil was based), in Bermuda
(where the Barracuda Tanker Company was registered), in
Liberia (where the ship was registered), or even in the
British courts. Since there is little uniformity in maritime
law throughout the world, the amount of recovery sought would
depend upon the country in which the action was filed.

The reverberations from the hulk on Cornwall's Seven
Stones Reef will be felt around the world for years. Lawyers,
courts, and the inadequacy of the law itself, indicate that
the actions growing out of the Torrey Canyon mishap could
outlast Charles Dicken's case of Jarndyce V. Jarndyce.

The Torrey Canyon fiasco made the entire world alert to
the need for more effective control of oil pollution of the
sea. That disaster made front page news everywhere, and
pulic alarm became focused upon the threat of oil pollution
from gigantic tankers. What most observers do not know is
that the greater hazard to marine environment comes from
oil spills which occur so frequently at terminals, shore
industrial plants, from leaking pipelines, refineries, docks,
offshore drilling sites, sunken tankers and natural oil
seeps.

The big tanker disasters, as dramatic as they are, are
not the major source of oil pollution. The most serious
pollution comes from the thousands of insidious incidents--
small ones, but preventable-- incidents of countless, minor
dumpings, and spills from thousands of tanker operations--
from emptying salt water ballast, pumping bilge water,
cleaning oil tanks, transferring and handling oil cargos.

In 1966, only 6 tankers were actually sunk at sea. This
amounts to less than 0.1 percent of the tankers in operation.
But of course when such a major tanker mishap occurs, the
results are devastating because of the great volumes of oil
that escape and persist in the entire marine environment.

Think of the pollution potential--about 1 out of every
5 vessels transports oil--oil amounting to 700 million tons
in 1966--about one-half of all goods transported at sea
from the standpoint of cargo weight. And these tankers are
busy. There were about 10,000 tanker visits to the United
States' ports alone in 1966. It is obvious that pipelines
are not delivering the entire crude oil requirements of the
industrial world.

Tankers are getting bigger, too, so there are greater
volumes of oil to spill. The average tanker size in 1955
was 16,000 tons. The tankers turned out lately average
76,000 tons, and some tankers being delivered now exceed
300,000 tons. Soon they will be one-half million tons. By
comparison, the Torrey Canyon carried 119,000 tons. Tankers
are so big now that the two officers of the new 1,010 foot
British tanker ESSO MERCIA were recently given bicycles to
make it easier to patrol the 166,820 ton vessel's decks.

A significant amount of the oil pollution of the sea
comes from offshore oil and gas operations. As of January 31,
1969, 7837 wells had been drilled by Federal oil, gas and
sulphur leasees on Outer Continental Shelf leases. The
most spectacular incident of oil pollution from offshore
drilling of course happened on January 18, 1969 in the
Santa Barbara Channel off the coast of Southern California.
Since that date countless experts and members of the general
public have indulged in arguments over the wisdom of drilling
in this area and whether stricter regulations and greater care
would have prevented the blowout.

Some observers of the Santa Barbara scene have urged the
Federal Government toward a policy of not offering leases in
any area where oil discharge of any kind would adversely
affect recreational beaches or fish and wildlife values.
Certainly this practice would eliminate the danger of pollution by oil, but it would also deny the U.S. Treasury untold
millions, indeed billions, of dollars each year. Such a
policy would also seriously hamper the country's effort to
locate and produce the oil experts say we must have to meet

the industrial commitment brought about by a soaring population that is expected to double in 50 - 60 years. In the
Santa Barbara Channel alone the Federal Government received
$603 million for 75 of the 110 lease blocks offered. It
seems quite unlikely under our stringent Federal budget of
today that neither the President of the United States nor
the Congress would seriously consider curtailing our offshore leasing program. Although safeguards of the highest
technology will be invoked, the spirit of the Outer Continental
Shelf Act is development[2].

The oil companies of this Nation are vitally concerned
with oil spills of every nature, and so are their shippers
and insurers. It is a matter of good economic sense. The
right company management policy forbids gambling with oil
pollution. As a result, in offshore drilling and transport
operations there are few companies with personnel or policies
so foolhardy that they compromise on inadequate equipment or
shoddy techniques. Prevention is the rule because spills
and blowouts may cost millions of dollars in lost oil, law
suits, and cleanup expenses. A further consideration for
avoiding oil pollution of the sea is of course the bad public
image that practice gains for the company and the hazardous
conditions it creates for employees.

Recognizing the hazards of polluting the sea with oil,
and the need to control it, let us examine what the Federal
Government's role has been and what it is likely to be in
the future, both internationally and at home.

The Convention on the Territorial Sea and the Contiguous
Zone, 1958, permits the United States to exercise within a
9-mile zone contiguous to its 3-mile territorial waters the
control necessary to prevent the infringement of its
"sanitary regulations within its territory or territorial
sea".

There is some debate in this country whether the United
States can enforce within the 12-mile band the measures now
available to us in the 3-mile territorial coastal waters under
the Oil Pollution Act, 1924, as amended.

Some maintain the Convention, by its ratification by
Congress, becomes enforceable; others say that specific

---

[2] 43 U.S.C.A. 1331.

enforcing statutes must be enacted to implement the
Conventions. This question has not been resolved, but the
Congress is considering legislation to expressly implement
the 1958 Convention and extend control of oil pollution to
the contiguous zone.

The United States Government has mobilizied all its
existing authority to cope with spills of oil and other
hazardous substances. We have developed a set of contingency
plans, tailored to various parts of our Nation, involving
all talent and equipment available to the U.S. Government
under existing laws. In 1968 the President ordered the most
effective system devised to discover and report a pollution
incident, stop the spread of oil, cleanup and dispose of the
pollutants, and institute available legal actions to recover
cleanups costs and enforce other federal statutes. The
plans will operate in our inland, coastal and territorial
waters, the contiguous zone and the high seas beyond this
zone, where there exists a threat to United States waters,
shores or continental shelf.

The International Convention for the Prevention of
Pollution of the Seas by Oil, 1954[3], as amended by the Con-
ference of Contracting Governments, is a direct approach
toward international control of oil pollution from vessels.
Thirty-eight contracting governments are members of this--
the only international agreement on oil. This Convention
is implemented in the United States by the Oil Pollution
Act of 1961, as amended. This Convention:

> Defines prohibited zone(offshore bands,
> 50 or 100 miles wide) in which discharge
> of oil are regulated.
>
> Requires logging of oil discharges and
> losses.
>
> Obliges signatory governments to promote
> the installation of oil-receiving facilities
> in their ports.
>
> Sets procedures for apprehension and
> prosecution of vessels which violate the
> provisions of the Convention.

---

[3] 33 U.S.C.A. 1001 et seq.

The U.S. Coast Guard enforces these provisions in the United States waters. The International Maritime Consulative Organization, a body of the United Nations, has study and recommending power in the area of oil pollution, but has no enforcement authority.

The Federal Water Pollution Control Act[4], as amended, is an important Congressional mandate in the prevention of water pollution from shore facilities. The Act requires the States to establish enforceable water quality standards applicable to interstate and coastal waters. These standards must be approved by the Secretary of the Interior, and must protect public health and welfare, and enhance the quality of the water. If a State fails to establish acceptable standards, the Secretary of the Interior is empowered to adopt such standards. Actually the standards of all states have been approved totally or in part, and continuous effort should be made by the Federal and State governments to keep the standard effective.

The water quality standards and their enforcement is one of the most effective methods of preventing pollution of a continuing nature, like the steady effluent from a coastal refinery. However, the time period between notice of a violation and its abatement is unreasonably long in the case of sudden, non-recurring pollution incidents.

The Oil Pollution Act, 1924[5], as amended, now makes unlawful, with some exceptions, the "grossly negligent" or "willful" discharge of oil from a vessel into U.S. navigable waters and adjoining shorelines. The Act applies to foreign and domestic vessels within our territorial sea and navigable inland waters and establishes both civil and criminal sanctions for violations. Except from liability are cases related to emergencies imperiling life or property, and unavoidable accidents, collisions, or stranding, and those cases where discharges are permitted by regulations established by the United States Government.

The 1924 Act was the first by Congress to specifically forbid pollution by oil discharges. In actual practice the law originally established a form of strict liability since

---

[4] 33 U.S.C. 466 et seq.
[5] 33 U.S.C. 431 et seq.

the only defenses applicable were that the discharge was an
emergency or an unavoidable accident. As previously indicated
prior to a 1966 Amendment, the U.S. Coast Guard used this
statute for sudden non-recurring violations rather than the
Federal Water Pollution Control Act. But in 1966 the Congress,
apparently believing that strict liability was too stern a
measure of conduct, amended the 1924 Act to require proof
that a discharge must be "willful" or "grossly negligent"
to be actionable. Since these terms are almost synonomous
legally, prosecutions have been rarely successful under the
1966 Amendment to the Oil Pollution Act of 1924. Congress
has been severely criticized for this step backward and in
this session is considering legislation to return some kind
of strict liability or to ordinary negligence.

The Rivers and Harbors Act of 1899[6], also known as the
Refuse Act provides:

> It shall not be lawful to throw, discharge, or
> deposit...any refuse matter of any kind or
> description whatever other than that flowing
> from streets and sewers and passing there-
> from in a liquid state into any navigable
> water of the United States... 33 USC 407 (1964ed.).

This statute was an attempt to consolidate four previous
Acts of Congress relating to specific kinds of refuse and
pollutants[7]. It is administered by the Corps of Engineers
and applies to both vessels and shore installations, and to
oil pollution in navigable waters which includes the coastal
zone within the three mile limit. A defendant convicted
under the Refuse Act may be fined $500 to $2,500 and be
imprisoned for not less than 30 days nor more than one year.
It may seem strange that informers are entitled to one-half
of the fine, but this provision must have been an afterthought
prompted by the difficulty in catching violators.

---

[6] Act of 1886 (24 Stat. 329); Act of 1888 (25 Stat. 209);
Act of 1890 (26 Stat. 453); Act of 1894 (28 Stat. 363).

[7] See 33 U.S.C. 466g(c)(5) requiring 180 days notice before
prosecution.

Although not originally intended to curb oil pollution, in interpreting the Refuse Act the United States Supreme Court has said that the history of the term "refuse of any kind or description" and the related legislation dealing with our free-flowing rivers forbids "a narrow, cramped reading of Section 13." U.S. v. Republic Steel Corp., 362 U.S. 482, 80 S. Ct. 884.

In a more recent decision the Supreme Court sustained a criminal prosecution where the defendant argued that commercially valuable gasoline accidentally discharged into Florida's St. John River did not constitute "refuse" under the 1899 Act. The Court said, in putting the defendant down:

> Oil is oil whether usable or not by industrial standards it has the same deleterious effect on waterways. In either case, its presence in our rivers and harbors is both a menace to navigation and a pollutant. U.S. v. Standards Oil Company, 384 U.S. 224; 86 S. Ct. 1427.

Under the Oil Pollution Act of 1924, as amended, a discharge of oil from a vessel under the terms of the statute must be removed by the person making the discharge. This, however, is assuming that the discharge was willful or grossly negligent. Discharges from other sources are not covered by existing law. If there is a failure to remove the discharged oil the Secretary of the Interior may proceed to remove that oil at the expense of the person creating the discharge. In addition there are criminal penalties of up to one year imprisonment or by a fine not exceeding $2,500, or both. A penalty of up to $10,000 may be invoked, and the same shall constitute a lien, against the boat or vessel. It is evident that these provisions fall woefully short of protecting the public interest when we consider the magnitude of disasters like the Torrey Canyon and the Santa Barbara Channel.

Pending legislation in the Congress of the United States is expected to make significant changes in the role of the Federal Government in preventing and controlling oil pollution of the sea. The House of Representatives has already passed, with but one dissenting vote H.R. 4148. Senate Bill No. 7 has been reported by the Senate Public Works Committee and there seem little doubt that by the end of the 1st Session

of the 91st Congress this bill will have passed in the Senate
and that an enrolled bill combining both H.R. 4148 and S. 7
will be sent to the President for signature.

One provision upon which there appears little disagree-
ment is that of amending the definition of discharges as it
now appears in the Oil Pollution Act of 1924. Similarly,
the final bill will also apply to discharges from both onshore
and offshore facilities.

Upon considering the House passed bill and S. 7 it is
readily apparent that the Congress intends to effectively
attack the matter of removing massive oil spills and blowouts
as well as discharges of a lesser nature. In the case of a
discharge from vessels S. 7 would hold the owner or operator
liable for the cost of removal to the extent of $14 million
or $125 per gross ton, whichever is less. Should the dis-
charge be willful or negligent the liability would be for
the entire cost of removal. The liability for removal under
the House passed bill would be the lesser of $100 per ton
or $10 million but only if the discharge was willful or
negligent.

With respect to oil discharges from onshore and offshore facilities, H.R. 4148 and S. 7 contain significant
differences and distinctions relating to matters of strict
liability and negligence, the amount of liability for removal
and the applicability of the legislation to discharges in
the contiguous zone and beyond.

It seems clear that oil pollution of the sea will continue
to be a National and International danger to our marine
environment throughout the foreseeable future. It poses an
ecological threat that is easily translated into economic
losses, as well as unmeasured social costs throughout the
entire world. The Executive Branch of the Federal Government has been endowed by the Congress with new and imaginative
authority to cope with increased oil pollution hazards in
the Nation's coastal zones. But oil, because of its migratory
nature, must be controlled beyond our traditional territorial
waters, not only by further Congressional action, but by
International treaties and conventions which will truly
establish the kind of liability that discourages oil pollution
at sea.

INDEX